T0334496

ANALYTICAL METHODS FOR BIOMASS CHARACTERIZATION AND CONVERSION

Emerging Issues in Analytical Chemistry

Series Editor
Brian F. Thomas

Co-published by Elsevier and RTI Press, the *Emerging Issues in Analytical Chemistry* series highlights contemporary challenges in health, environmental, and forensic sciences being addressed by novel analytical chemistry approaches, methods, or instrumentation. Each volume is available as an e-book, on Elsevier's ScienceDirect, and via print. Series editor Dr. Brian F. Thomas continuously identifies volume authors and topics; areas of current interest include identification of tobacco product content prompted by regulations of the Family Tobacco Control Act, constituents and use characteristics of e-cigarettes and vaporizers, analysis of the synthetic cannabinoids and cathinones proliferating on the illicit market, medication compliance and prescription pain killer use and diversion, and environmental exposure to chemicals such as phthalates, endocrine disrupters, and flame retardants. Novel analytical methods and approaches are also highlighted, such as ultraperformance convergence chromatography, ion mobility, in silico chemoinformatics, and metallomics. By highlighting analytical innovations and new information, this series furthers our understanding of chemicals, exposures, and societal consequences.

2016 *The Analytical Chemistry of Cannabis: Quality Assessment, Assurance, and Regulation of Medicinal Marijuana and Cannabinoid Preparations*
Brian F. Thomas and Mahmoud ElSohly

2016 *Exercise, Sport, and Bioanalytical Chemistry: Principles and Practice*
Anthony C. Hackney

2016 *Analytical Chemistry for Assessing Medication Adherence*
Sangeeta Tanna and Graham Lawson

2017 *Sustainable Shale Oil and Gas: Analytical Chemistry, Geochemistry, and Biochemistry Methods*
Vikram Rao and Rob Knight

2017 *Analytical Assessment of e-Cigarettes: From Contents to Chemical and Particle Exposure Profiles*
Konstantinos E. Farsalinos, I. Gene Gillman, Stephen S. Hecht, Riccardo Polosa, and Jonathan Thornburg

2018 *Doping, Performance-Enhancing Drugs, and Hormones in Sport: Mechanisms of Action and Methods of Detection*
Anthony C. Hackney

2018 *Inhaled Pharmaceutical Product Development Perspectives: Challenges and Opportunities*
Anthony J. Hickey

2018 *Detection of Drugs and Their Metabolites in Oral Fluid*
Robert M. White, Sr., and Christine M. Moore

ANALYTICAL METHODS FOR BIOMASS CHARACTERIZATION AND CONVERSION

DAVID C. DAYTON
RTI International, Research Triangle Park, NC, United States

THOMAS D. FOUST
National Renewable Energy Laboratory (NREL), Golden, CO, United States

Elsevier
Radarweg 29, PO Box 211, 1000 AE Amsterdam, Netherlands
The Boulevard, Langford Lane, Kidlington, Oxford OX5 1GB, United Kingdom
50 Hampshire Street, 5th Floor, Cambridge, MA 02139, United States

Library of Congress Cataloging-in-Publication Data
A catalog record for this book is available from the Library of Congress

British Library Cataloguing-in-Publication Data
A catalogue record for this book is available from the British Library

ISBN: 978-0-12-815605-6

For information on all Elsevier publications
visit our website at https://www.elsevier.com/books-and-journals

Publisher: Susan Dennis
Acquisition Editor: Kathryn Eryilmaz
Editorial Project Manager: Devlin Person
Production Project Manager: Surya Narayanan Jayachandran
Cover Designer: Matthew Limbert and Dayle G. Johnson

Typeset by SPi Global, India

Dedication

David Dayton would like to dedicate this book to several individuals who made significant impressions on a young research scientist at the National Renewable Energy Laboratory (NREL). Dr. Thomas Milne was my mentor in the early days of my career and taught me that focusing on the science was most important. Dr. Robert Evans took a chance on me and gave me my first job in biofuels at NREL. Dr. Helena Chum's tireless efforts at NREL to sustain a vibrant biofuels research center had a profound impact on me in the early years, as did many technical and philosophical discussions with Drs. Ralph Overend and Richard Bain. Rich became a confidant during my years at NREL, and his chemical engineering perspective complemented my chemistry background. Collectively, their passion for research and dedication to the mission of producing renewable fuels for a more sustainable energy future was infectious and kickstarted my career. Countless other co-workers at NREL and RTI and colleagues in the biofuels community are too numerous to mention individually, but I have thoroughly enjoyed sharing with them the common goal of making alternative biofuels the norm.

Contents

Foreword

The World has grown highly dependent on petroleum to power transportation and produce a wide range of chemicals and other products, all of which are so important to attaining our current standard of living. Natural gas and coal are the other primary fossil resources on which we have relied heavily for production of electricity as well as for home and building heating and manufacture of various products. However, despite the current low cost and abundance of these fossil resources due to advances in horizontal drilling, fracking, and other exploration tools, it must be kept in mind that fossil resources are ultimately finite. And as the standards of living improve as less developed countries seek to follow the examples of the developed world, the drain on these resources will grow, moving them closer to scarcity. Beyond dealing with ultimate scarcity of fossil resources that would imperil comfortable lifestyles, reliance on these resources has serious consequences, including strategic vulnerability, balance of trade issues, and price volatility, that have been evident in the not too distant past by such examples as the OPEC oil embargo. Thus, at some point, we need to transition to sustainable sources from which we can provide energy to drive advanced economies as well as derive the array of products so vital to maintaining our lifestyles.

A far more hideous and urgent, but so far less obvious consequence of such widespread use of fossil energy is global climate change due to the build-up of carbon dioxide in the atmosphere as fossil carbon is moved from below the ground into the air by burning fossil resources. Because carbon dioxide is a powerful greenhouse gas that traps and reradiates infrared radiation back to the Earth, its build-up will rapidly produce major changes in our climate with resulting dire consequences. Yet because such detrimental effects will take a few decades to become more apparent and about a century to be fully recognized, many deny it will happen, while others who recognize the consequences have not taken meaningful action. The result is that little has been done to curb this ultimately disastrous outcome.

So what could be done to keep this high standard of living and reduce carbon dioxide emissions? The answer is obvious: rapidly transition to sustainable and other low carbon energy sources such as solar photovoltaics and heating, wind, geothermal, and nuclear. Biomass is among these options in that it can be regenerated by capturing the carbon dioxide emitted when it is burned in a closed carbon cycle that results in low net carbon emissions as long as little if any fossil energy is used to grow, harvest, transport, and convert biomass to fuels or for the distribution and use of resulting fuels. Although this may seem like a tall order, it is eminently feasible if the overall cycle is carefully tended to. Structural plant biomass such as grasses, wood, and agricultural and forestry residues are particularly well suited to following this cycle in that they can be grown and converted to fuels with minimal fossil inputs, and their use need not interfere with feeding the world. Although biomass utilization is often downplayed due to various concerns, this resource is uniquely suited for production of liquid and gaseous fuels that we so much prefer, as well as to manufacture chemicals that no other sustainable option can match. For example, although fuel cells and batteries could make sustainable electricity from wind and photovoltaics viable for powering passenger cars and other light duty vehicles, a daunting amount of materials will be required to fully electrify these sectors. In addition, while fuels from biomass, that is to say biofuels, can meld into the existing infrastructure, substantial changes will be necessary to support the transition to batteries and fuel cells. Even if full electrification is possible, only biofuels are well suited to fuel aircraft or long-haul trucking, at least until substantially more energy-dense and low-cost battery and fuel cell power options are developed in the likely distant future. Transitioning to biofuels can help reduce build-up of carbon dioxide in the meantime.

So how can biomass, which is a solid, sustainably meet our strong preference for liquid and gaseous fuels? The answer: through biological or thermochemical biomass conversion or hybrid combinations of the two. First, the good news. The large number of biomass sources and types and options for their conversion greatly enhance the possibility for successful development and commercialization of cost-competitive, sustainable production of biofuels. However, the bad news is that pursuing so many pathways can dilute progress on the most promising, and unfulfilled promises about routes with little chance of success colors perception and support for those that can make a difference. In any event, understanding and advancing biomass and its conversion technologies requires following valid protocols for characterizing biomass resources and products to validate the potential for competitiveness and desirable environmental attributes. And most of these protocols are very demanding in terms of both time and care.

Against this background, this book presents insights gained by two experienced contributors to the biomass field that should aid those new to the field to gain perspectives on biomass conversion and characterization technologies. The information presented can also be valuable to more experienced practitioners. However, because the field is so broad, multifaceted, and complex, only a limited view can be presented in a single book that seeks to cover such a broad area. It is also important to keep in mind that condensing so much information requires simplification that masks complex details. In addition, with so many technologies under development, only selected examples can be covered, with the result that promising options can be missed. Thus, it is vital for the reader to build from the information and references provided as well as independently dig into the literature more deeply to develop a more complete picture of this complex field, and to seek more details from other sources for areas in which this book invites interest.

Charles E. Wyman is the Ford Motor Company Chair in Environmental Engineering in the Center for Environmental Research and Technology and a Distinguished Professor in the Chemical and Environmental Engineering Department at the University of California Riverside. He is also a Team Leader in the Center for Bioenergy Innovation at Oak Ridge National Laboratory. Professor Wyman has devoted most of his career to leading advancement of technology for conversion of cellulosic biomass to ethanol and other products. His research has focused on pretreatment, enzymatic hydrolysis, and dehydration of cellulosic biomass to produce reactive intermediates for conversion to fuels and chemicals. He has published about 200 technical papers, contributed 40 chapters and edited volumes, and has 21 patents. In addition to his role as a founder and Co-Editor-in-Chief of *Biotechnology for Biofuels*, he recently founded Vertimass LLC, a company devoted to commercializing a novel catalytic technology for simple one-step conversion of ethanol to fungible gasoline, diesel, and jet fuel blendstocks.

Preface

This book is intended to be a primer or introduction to information we view as critical to understanding the increasing role for biomass as concerns about energy security, environmental quality, and global climate change drive the transition to renewable and ultimately carbon neutral or negative energy resources. It gives an overview of the activities in biofuels research and development that we have experienced during our careers and how analytical methods for characterizing biomass, intermediates, and products have impacted scientific understanding and technology development. The focus is on biomass as a resource for developing biofuels technologies, but opportunities for bio-products and bio-power production are also discussed.

The book is organized to follow the process of collecting and converting biomass into advanced biofuels and bio-products in an economically competitive, integrated process that is environmentally sustainable. Chapter 1 provides definitions and background information to set the context for subsequent chapters. Chapter 2, "Biomass Characterization," describes the physical and chemical properties of biomass and techniques for measuring desired properties. Often, how we characterize biomass is a function of which technology is being considered for converting it into product. Chapter 3 reviews the available and developing technologies, biomass feedstocks best suited for specific conversion processes, and intermediates and products that can be produced by different technology options. Chapters 4–6 summarize the analytical methods used for probing biochemical, thermochemical, and hybrid conversion technologies, respectively. Detailed characterization of all process streams (inputs and outputs) supports process development and optimization.

Chapter 7 summarizes the economics of selected integrated advanced biofuels processes, and Chapter 8 details the methods used for determining the environmental impacts of these technologies from a life-cycle perspective. Chapters 9–12 introduce the technology needs and the non-technical regulatory and policy challenges associated with several commercial end-use applications of biomass and biofuels. Chapter 9 takes an in-depth view at an alternative biofuels approach for leveraging the high oxygen content of biomass to produce oxygenates that have improved combustion behavior in engines modified to take advantage of the alternative fuel. The focus of Chapter 10 is on the challenges with producing and certifying sustainable alternative jet fuels to meet commercial aviation industry commitments for reducing the carbon footprint of air travel. Chapter 11 follows with a description of biochemical and thermochemical conversion routes for producing higher value bio-based products, chemicals, and materials to improve the economics of developing biorefineries. Chapter 12, "Waste to Energy," looks at the challenges and opportunities of using a ubiquitous low-cost feedstock for bioenergy production.

Biofuels are viewed as an important part of the solution for reducing greenhouse gas emissions to help slow or reverse the impact of transportation on global climate change. We would be remiss if we did not devote some discussion to other potential carbon-neutral or even carbon-negative options. Chapter 13 summarizes products with potential for carbon sequestration and more futuristic technologies that may eventually use CO_2 as a feedstock for fuels, chemicals, and polymers. Chapter 14 concludes with a discussion of some of the challenges we see remaining for the future of biofuels, bio-products, and bio-power.

Each chapter provides an overview with selected references that reinforce and support specific examples to introduce the biofuels beginner to key aspects of the field. The Additional Reading selections provide references to what we feel are the seminal biofuels publications encountered during our careers and will give biofuels researchers, policy makers, and businesses a solid starting point for delving deeper into topics or research areas in which they are most interested. For beginner or expert, we hope the book gives existing and future biofuels enthusiasts a firm foundation to support their own scientific journey toward a sustainable energy future.

Acknowledgments

We want to acknowledge the great help and support of the people at RTI International in North Carolina, especially Drs. Gerald T. Pollard and Brian F. Thomas, without whom this work would not be possible. Many thanks also to Dayle G. Johnson of RTI for the cover design. We also want to acknowledge Dr. Anna Wetterberg and RTI Press for supporting our efforts with editing and graphics development. Dr. Dayton would like to thank the RTI Fellows Program for financial support for completing this project.

The authors are indebted to the Department of Energy, Office of Energy Efficiency and Renewable Energy for financial support of biofuels research and development throughout our careers, starting with the Office of Biomass Power and Office of Biofuels in the 1990s, the consolidated Office of Biomass Programs in the 2000s, and the current Bioenergy Technologies Office.

We also want to thank all of our contributors, who provided their expertise for additional insights into specific topics, and all of our colleagues with whom we interacted during this project.

CHAPTER ONE

Introduction

Historical Use of Energy Resources

The Earth is an optimized system for collecting and storing solar energy. Photosynthesis is the process by which solar energy is used to convert carbon dioxide (CO_2) and water (H_2O) into cellulose, hemicellulose, and lignin, the three biopolymers that constitute plants, or what we call biomass. Cellulose and hemicellulose are polysaccharides for storing energy. Lignin is the glue that holds it all together and provides structural integrity.

Fossil fuels started out as biomass. Terrestrial biomass is plants and trees that grew on land, fell to the ground, decayed over time, and provided for the growth of new vegetation. After millions of years, layers of plant matter formed by this cycle were buried beneath the surface of the planet. As the layers became buried ever deeper, the increased pressure and temperature forced out the oxygen, forming a carbon-rich deposit that we call coal.

Aquatic plants and animals had the same fate over geologic time scales. They died, sank to the bottom of prehistoric oceans, and became buried in mud and sediment, where the pressure and temperature transformed them into crude oil deposits in sedimentary rock formations. The main difference between aquatic and terrestrial biomass is lignin content. Aquatic biomass growing in and on water contains less lignin because it requires less structural support than terrestrial biomass that continually must overcome gravity to remain standing. Consequently, aquatic biomass forms crude oil over geologic timescales while lignin-containing terrestrial biomass forms coal.

Biomass is among the oldest sources of stored energy. When burned, it provides light and heat for warmth and cooking. Fig. 1.1 illustrates how energy resources for heating in the United Kingdom have changed over the course of history. Wood was the predominant fuel source in early civilization giving way to coal over time, probably because of its greater energy density. Only recently has petroleum, natural gas and electricity become the predominant energy resources for heating. Energy consumption in the United Kingdom is correlated with population growth and increasing

Analytical Methods for Biomass Characterization and Conversion
https://doi.org/10.1016/B978-0-12-815605-6.00001-9

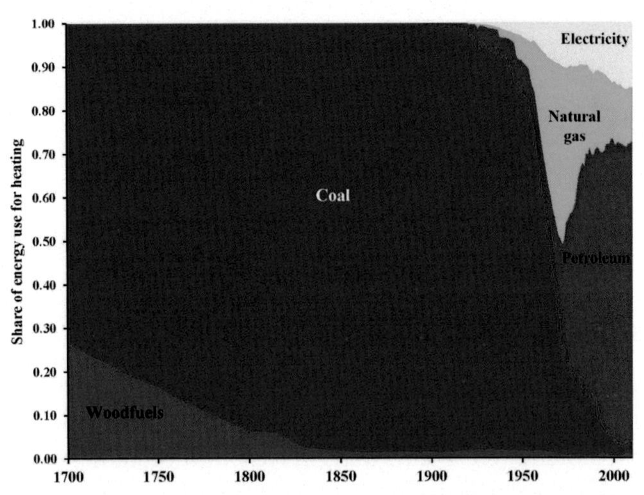

Fig. 1.1 Share of energy consumption for domestic and industrial heating in the United Kingdom from 1700 to 2010. *(Adapted from Fouquet R. The slow search for solutions: lessons from historical energy transitions by sector and service.* Energy Policy *2010;38 (11):6586-6596.)*

prosperity. These trends hold for the developed world, though the timescale differs among countries and regions. Current energy consumption in the developing world can be quite different and depends strongly on available resources.

Access to energy was a key element in the foundation and maintenance of civilization, but societal advancements were enabled by shifts toward specialized and productive economic activities fueled by more efficient use of available energy resources. The United States Energy Information Agency tracks historical energy production and use, and forecasts demand according to energy resource and sector. Fig. 1.2 shows energy usage in the United States from 1776 to 2012 as a function of resource type. Per capita consumption was low in colonial days, with wood as the primary resource. Wood utilization peaked around 1860, and then coal increased to become primary by 1900. Petroleum and natural gas consumption were increasing rapidly by 1950 and overtook coal as the primary source by 1960. Availability of renewable energy resources other than wood has historically been very low. Interest increased in the 1970s, particularly after the oil embargo in 1973. The increase in renewables production since 2005 is due to the exponential growth in the corn ethanol industry and the cost reductions associated with wind and solar technology. Clearly, energy consumption in the United States correlates with increasing population, and access to low-cost, abundant energy is linked to economic prosperity.

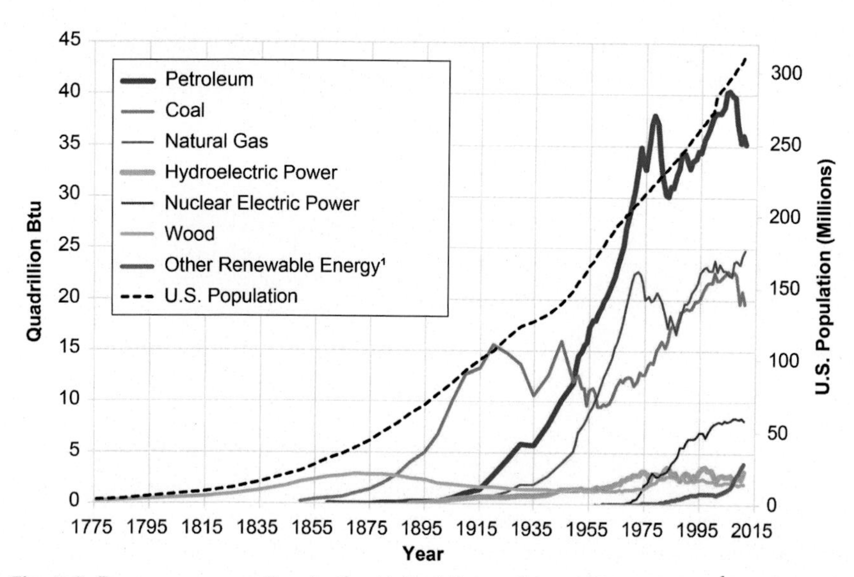

Fig. 1.2 Energy consumption in the United States from 1776 to 2012. [1]Geothermal, Solar/PV, wind, waste, and biofuels. *(Source: US Energy Information Administration Energy Perspectives "Annual Energy Review" and "Monthly Energy Review" Tables 1.3, 10.1, and E1).*

The historical trends in energy use for transportation are quite different. Fig. 1.3 shows the percentage of various energy resources in the United Kingdom from 1700 to 2010. Early transportation was by foot and horse on land, sailing and rowing on water. The development of the steam engine led to coal-powered locomotives and ships in the early 19th century. The discovery of oil in Pennsylvania in 1859 catalyzed the development of the internal combustion engine for vehicle transportation toward the end of the century. Subsequent discoveries of oil in Texas, Oklahoma, and the Gulf Coast of the United States produced the modern petroleum industry. Crude oil is now a global commodity, recovered all over the world. According to the United States Energy Information Agency, the United States, Saudi Arabia, Russia, Canada, and China were the top five oil producing countries in 2018. Petroleum distillates are the dominant fuels for transportation and the basis for an extensive petrochemical industry. Biomass has been much less used as a transportation fuel.

The hydrocarbon landscape changed dramatically in the United States as a new energy revolution was emerging, due in large part to advances in horizontal drilling and hydraulic fracturing to extract untapped oil and gas from shale formations. The United States was importing just over 9 million barrels per day (MMbpd) of oil in 2010, which was down from 10 MMbpd in 2005 caused largely by the global recession. In May 2015, imports were down to

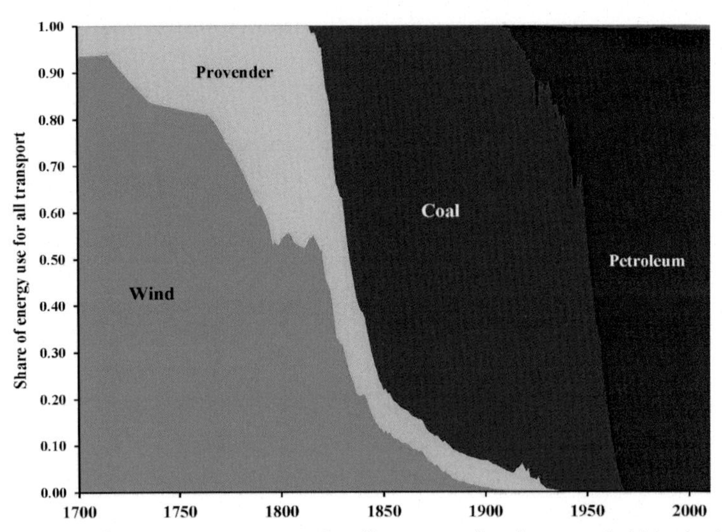

Fig. 1.3 Share of energy consumption for all transport (land, sea, and air) in the United Kingdom from 1700 to 2010. *(Adapted from Fouquet R. The slow search for solutions: lessons from historical energy transitions by sector and service.* Energy Policy *2010;38 (11):6586–6596.)*

6.9 MMbpd because of a drop in consumption and an increase in United States shale oil production.[1] In April 2015 shale oil production was 9.4 MMbpd, nearly double the capacity in 2008 (5.0 MMbpd). For comparison, the corn ethanol industry production capacity was 0.25 MMbpd in 2005 and 0.95 by 2015.[2, 3]

In late 2014, oil prices dropped steeply, as shown in Fig. 1.4. US annual production had been rising at the rate of 1 MMbpd per year for over 2 years to balance the demand caused by shortfalls across the globe—in Libya, for example. But the United States kept adding capacity at the same rate while global production started to slowly increase. Concurrently, petroleum consumption in China slowed, eventually creating a global oversupply of crude. Also, one of the largest global crude oil producers, Saudi Arabia, did not reduce production, hoping to profit from the higher prices. The confluence of these events caused crude oil prices to drop by 55% in a matter of months.

Shale production in the United States is somewhat like a factory, in that wells can be put on stream or removed from production quickly. Consequently, US production can be more responsive to price fluctuations in the global crude oil market than the Organization of the Petroleum Exporting Countries (OPEC). The low crude prices did not last for long; recent prices have been in the $35–$60 per barrel range.

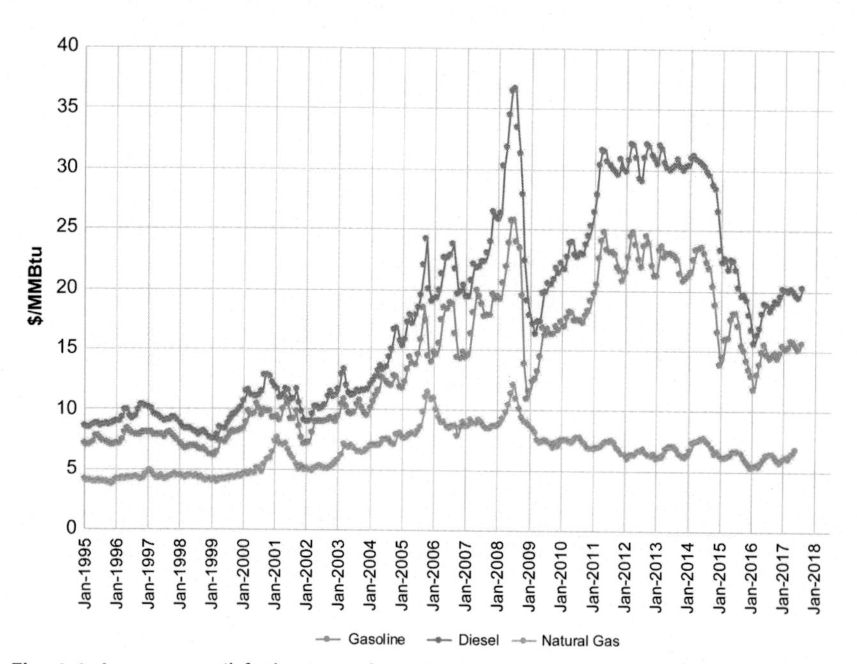

Fig. 1.4 Average retail fuel prices of gasoline, diesel, and natural gas on an energy content basis. Gasoline is an average retail price of all grades and all formulations, diesel is the retail price of No. 2 diesel, and natural gas is the US price of natural gas sold to commercial consumers. *(Based on data from the US Department of Energy's Energy Information Agency.)*

Prior to about 2010, natural gas prices had been unpredictable and usually trended with liquid petroleum fuels, as shown in Fig. 1.4. In 2005, natural gas had surpassed $12 per MMBtu (million British thermal units) for brief periods, though the baseline price had been closer to $6. The volatility was a major risk factor that prohibited increased natural gas use for energy production. But after about 2010, the increased production of shale gas drove the price down and reduced the volatility, even as petroleum prices remained volatile. The Energy Information Administration forecasts natural gas prices to stay low to moderate for decades.[1]

This assurance in natural gas pricing is driving two types of behavior. One is substitution of diesel in short-haul trucks with compressed natural gas (CNG) and in long-haul trucks with liquefied natural gas (LNG).[4] The other is the conversion of natural gas into chemicals and fuels previously sourced from oil.[5] The driver behind these two trends has been primarily economic.

The second area—gas conversion to drop-in fuels—is still at an early stage. Sasol and Shell initially announced plans to build gas-to-liquids plants in Louisiana using Fischer-Tropsch synthesis to produce largely diesel fuel. Although long-term natural gas prices are not in doubt, the uncertainty in the price of oil appears to be giving these companies pause for making substantial capital investments. In contrast, at least half a dozen world-scale methanol plants are well into planning and construction.

In the transportation sector, the development of engines and the fuels they run on have been intimately coupled since the advent of the automobile industry. There is no better example than Henry Ford's Model T and claims that is was the first "flex-fuel" vehicle. A quick search of the Internet will reveal that Ford did not intend the Model T engine to run on ethanol, but the low compression ratio did make it possible to use a wide range of fuels. The fuel distribution infrastructure at that time was not well developed, so, out of necessity, consumers were somewhat unselective when it came to the hydrocarbons they put in their tanks. As engines advanced, so did the fuels to meet the demand for better performance and higher efficiency.

Higher compression ratios for more power and efficiency required higher quality fuel. Uncontrolled combustion affected the air-fuel ratio and upset ignition timing. Preignition caused engine knock, which reduced power and increased emissions. Knock was eliminated by higher quality fuels, namely higher octane. High performance engines requiring high octane fuel was essential for the developing aviation industry.

Energy and the Environment

Higher octane is achieved by adjusting the petroleum refining process to optimize the hydrocarbon blend or by using additives. In the 1920s, the refining industry was in its infancy, so low-cost additives were chosen. One of the first was tetraethyl lead. Even at the time, the neurotoxic effect of lead was well known. Adding it to gasoline made handling the fuel more dangerous, and the impact of lead emissions on air and soil quality became a much greater concern over time.

Other air pollutants such as oxides of nitrogen (NO_x) and sulfur (SO_x), particulate matter (PM) or soot, unburned hydrocarbons, and carbon monoxide became environmental issues as automobile transportation increased. The consequences of these pollutants can be long-term environmental issues and health effects. NO_x and SO_x cause acid rain, which leads to deforestation. Unburned hydrocarbons and carbon monoxide emissions cause unhealthy levels of ground level ozone, which increases respiratory problems and aggravates lung diseases like asthma and bronchitis. Small PM and NO_x emissions have been implicated in lung and heart disease.

In 1970, the United States Congress passed the Clean Air Act to effectively reduce the environmental and health effects posed by six pollutants: PM, ground level ozone, carbon monoxide, NO_x, SO_x, and lead. Reducing emissions in the transportation sector required changes in all aspects of the value chain. Automobile manufacturers were required to build cleaner engines, petroleum refiners were required to produce cleaner burning fuels, and passenger vehicle inspection and maintenance programs were required in areas with acknowledged pollution problems. This not only directly reduced lead emissions but also eliminated a poison for the three-way catalysts that were developed to reduce NO_x, CO, and unburned hydrocarbons in tailpipe emissions. Leaded gasoline in on-road vehicles has now been essentially phased out across the world.

As lead began to be phased out in the mid-1970s, alternative solutions for raising octane were sought. Advances in petroleum refining led to the development of reformulated unleaded gasoline. Early versions had lower octane than leaded predecessors. Oxygenated additives such as methyl *tert*-butyl ether (MTBE) and ethanol served two purposes: to increase octane rating and reduce CO and unburned hydrocarbon emissions.

Groundwater contamination by leaking underground fuel storage tanks lead to higher than desired levels of MTBE in drinking water. MTBE is not classified as a human carcinogen, but even low levels give water an unpleasant taste. The financial risks from impending litigation to remove MTBE from contaminated soil and groundwater and the increasing regulation for reducing MTBE in drinking water forced its phase-out as a gasoline additive. Consequently, ethanol has become the predominant oxygenated additive—just in time to capitalize on the surge in the renewable fuels industry. Low corn prices and high market demand for oxygenates in reformulated gasoline provided the economic incentive for rapid industry growth. The principal avenue of growth was through adding ethanol, at first up to 10% by volume, and enabling supply by subsidizing the industry if needed. As petroleum prices rose, blending ethanol in gasoline actually reduced transportation fuel cost in some areas.

Ethanol has a higher octane rating (109) than gasoline (85–94) but contains only two-thirds of the energy per unit volume, so it delivers fewer miles per gallon. According to the US Environmental Protection Agency (EPA), a 10% ethanol blend delivers 3–4% fewer miles per gallon than nonblended gasoline. As the ethanol content increases, the mileage penalty increases. Flex-fuel vehicles designed to run on 51–85% ethanol get 15–30% lower mileage than vehicles using regular gasoline with 10% ethanol.

However, higher ethanol blends support higher compression ratios and more aggressive ignition timing without the risk of knocking in high-performance turbocharged engines. Also, ethanol has a higher heat of evaporation than gasoline, so fuel injection causes substantial in-cylinder evaporative cooling, which raises the effective octane rating of high-ethanol blends even further. The cooler combustion temperatures also result in decreased NO_x emissions. Therefore, higher efficiency engines are required to take full advantage of the potential environmental benefits of increased ethanol use.

An increasing number of passenger vehicles run on diesel instead of gasoline, and most long- and short-haul trucking is done with diesel power. A higher compression ratio combined with the intrinsically higher energy content of diesel increases fuel efficiency by about 30% over gasoline engines of like size.

When a diesel is operated under slightly fuel lean conditions, the combustion temperature is maximized, so it performs better, especially with respect to fuel economy. This condition is defined as air in excess of the stoichiometric amount required to completely combust the fuel. Less unburnt fuel also reduces PM emissions. However, the excess air and higher combustion temperature causes higher NO_x emissions. Consequently, minimizing diesel engine emissions (PM and NO_x) cannot be accomplished simply by optimizing engine operating parameters, because lower PM is favored under lean burn conditions while lower NO_x is favored under fuel rich conditions. Some form of exhaust gas treatment is required.

Exhaust gas treatment with three-way catalytic converters is used to control NO_x, CO, and unburned hydrocarbons in gasoline powered vehicles. However, this technology is not applicable for diesel engines because of the soot formation and higher NO_x concentrations. In diesel exhaust, PM is generally minimized with lean burn conditions, but it is also relatively simply captured in filters.

NO_x reduction in diesel engine exhaust is another challenge. Reducing the combustion temperature by exhaust gas recycling will reduce NO_x, but this reduces power and fuel efficiency. Selective catalytic reduction (SCR) can also be used, much like the technology in electric power plants. Reductants such as ammonia or urea can be injected into the exhaust stream to react with NO and NO_2 over a catalyst to produce N_2; however, this is relatively expensive, and careful control is needed to avoid ammonia slip.

Another developing technology is the lean NO_x trap (LNT). This is a two-step process where the first step involves absorbing NO_x onto a catalyst surface during lean burn conditions. This is continued until the surface is saturated, leading to the second step, which regenerates the catalyst during short periods of fuel rich engine operation where the absorbed NO_x is catalytically converting to N_2 and CO_2. The catalyst material can be expensive, and the capacity of the LNT is limited, so "high NO_x emissions events" can occur when the engine operates under heavy load. However, LNT does not reduce engine efficiency except during the fuel rich catalyst regeneration periods, and additional tanks for reductant are not needed, so it is a cheaper option than SCR.

Changing the fuel composition is another strategy for reducing air pollutants. Reducing the sulfur content of diesel to manage SO_x is the most pertinent example. The US EPA began regulating the sulfur content of diesel transportation fuel in 1993. It has been decreased from 5000 to 500 ppm (low sulfur diesel, LSD) and in 2010 to 15 ppm (ultralow sulfur diesel, ULSD). The European Union was on a similar trajectory; as of 2007, the maximum was 15 ppm. Petroleum refiners had to modify hydrotreating unit operations to increase hydrodesulfurization to meet the ULSD specifications. Significant improvement in air quality is gained by using ULSD for cleaner burning diesel engines in vehicles with advanced emissions control devices.

The environmental and health effects associated with the criteria pollutants are short term in contrast to the impact of CO_2 emissions on global climate change. Scientists agree that consumption of fossil fuels by humankind has accelerated the increase in CO_2 emissions, associated with an increasing number of severe weather events and drought. Mitigating environmental effects from trace elements such as sulfur and nitrogen in petroleum has been successfully managed, but carbon is the most abundant element in petroleum (75 mol% and 84 wt%). As a result, approximately 4.6 million tons of CO_2 are emitted from a typical passenger vehicle each year. Multiply that figure by the number of vehicles (approximately 270 million in the United States), and CO_2 emissions from transportation are not trivial. The EPA estimated that the total greenhouse gas emissions in 2017 was 6456.7 million tonnes (metric tons), with 29% generated by transportation.[6] Capturing this amount of CO_2 seems impractical. The sensible ways to reduce the amount of fossil CO_2 emissions from transportation are to reduce the amount of fuel used and to substitute existing fossil fuels with lower-carbon options.

There are two options to reduce fuel consumption. One is to drive fewer miles. Public transportation, carpooling, alternate modes of transport (bicycles, for example), and telecommuting can be done by individuals. Availability of these societal options is strongly dependent on regional infrastructure and population density. The other is to drive more miles using the same amount of fuel or less. Developing more fuel-efficient vehicles by using lighter weight construction materials and higher efficiency engines is dependent on technological innovation and adaptation in the automotive industry. A highly visible example is the hybrid or all-electric vehicle. Any changes to vehicles must be done without compromising safety and performance, at a price point that the market will accept. Of course, the consumer ultimately has the choice, and one size does not fit all. The likely scenario is that both options will be adopted in some form.

In response to the Arab Oil Embargo, the US government established the Corporate Average Fuel Economy (CAFE) standards in 1975 to reduce transportation fuel consumption by increasing the fuel economy of cars and light trucks (light duty vehicles). The Department of Transportation's National Highway Traffic Safety Administration (NHTSA) administers the program and sets the standards for the maximum feasible average fuel economy that a given automobile manufacturer can achieve in a given model year. Compliance is based on the average fuel economy of the total number of domestic passenger cars, imported passenger cars, and light trucks sold in the United States by a manufacturer. The standard (expected mileage) for each of the three categories is compared to actual mileage. If the actual is higher than the standard, the manufacturer gains a credit; if lower, the shortfall must be recouped.

Fig. 1.5 shows the standards and achieved mileages of the three categories of vehicle for the model years 2004–16. Important to note is the general trend for increasing efficiency over time. Actual efficiency increased by 10 miles per gallon (mpg), from 28 to 38 for domestic and imported cars. Light trucks showed the same trend, though at a lower level than cars. Actual efficiency for domestic and imported cars was uniformly higher than the standards. Note that from 2004 to 2010, standards for domestic and imported cars were flat at 27.5 mpg.

The Energy Independence and Security Act of 2007 required the CAFE standards for cars to increase to 35 mpg by 2020, starting with model year 2011. As shown in the graph, these targets were already met in 2016, at

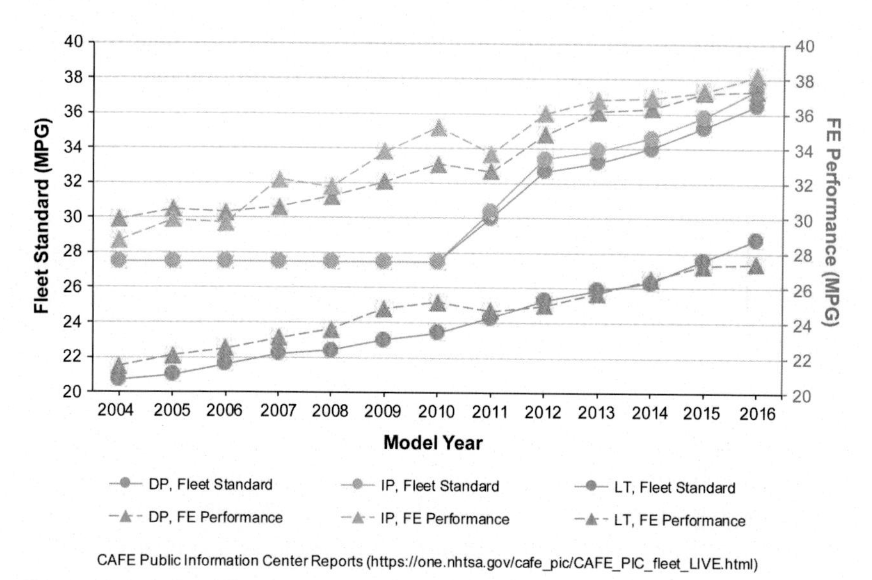

Fig. 1.5 Actual fuel efficiency compared to CAFE fleet standard for domestic cars (DP), imported cars (IP), and light trucks (LT).

38 mpg. This prompted the call for an even more aggressive target of 54.5 mpg for 2025 and beyond. That is the average for a given vehicle year. There will be considerable spread across the range of cars, but the trend of greater fuel efficiency is obvious.

Biomass and Biofuels

Reducing the greenhouse gas emissions from transportation can also be achieved by replacing conventional fossil fuels with more sustainable, renewable fuels. The Energy Policy Act of 2005 (EPAct 2005) was signed into law "to ensure jobs for our future with secure, affordable, and reliable energy." The objective was to provide market pull for developing a robust biofuels industry to diversify the US transportation fuel supply by achieving greater use of "homegrown" renewable fuels to increase energy security and improve air quality. Biofuels were seen as an important element for achieving these goals, with the added benefits of creating jobs in rural communities and providing more fuel choices for consumers. The Renewable Fuel Standard (RFS) was created to reduce the quantity of fossil fuels used for transportation by establishing annual volume targets for renewable fuel production between 2006 and 2022. The RFS was devised with two national objectives in mind: to reduce imports of petroleum and to reduce the greenhouse gas emissions associated with transportation.[7] The RFS program was enacted into law under the EPAct 2005 and established the first renewable fuel volume mandate in the United States. The Energy Independence and Security Act (EISA) of 2007 expanded the RFS program (RFS2) to include four unique categories of renewables: cellulosic biofuel, biomass-based diesel, advanced biofuel, and total renewable fuel.

Organic matter that can be converted into energy is the simplest definition of biomass. However, biofuels legislation (EPAct 2005 and EISA 2007) defined it as organic matter that is available on a renewable basis. The major types for energy conversion are food crops, agriculture wastes and residues, woody biomass, wood wastes and forestry residues, animal wastes (manure), municipal wastes, and aquatic plants such as micro- and macroalgae.

Defining biomass as renewable means that it will be available perpetually, if managed properly, to recycle carbon in the biosphere. Renewable resources can be used to produce biofuels, bio-products, and bio-power that offset the carbon from burning non-renewable sources, such as fossil fuels. In this context, the concept of sustainability dictates that as biomass is used, it needs to be replaced to maintain a global carbon balance. Therefore, carbon accumulation in the atmosphere can be slowed or reversed if sustainable biomass resources are efficiently and economically converted into biofuels, bio-products, and bio-power that displace the consumption of fossil fuels, which emit carbon that has been sequestered beneath the Earth's surface for millions of years.

Global demand for biomass is expected to double by 2050.[8] Much of this growth will occur from agricultural intensification to provide a sustainable food supply for a growing global population with increasing standards of living.[9] Bioenergy demand could put additional pressure on arable land use. The International Renewable Energy Agency describes a roadmap for global sustainable bioenergy production to meet increasing energy demand while curbing climate change.[10] But is there enough biomass available, now and in the future, to make a difference?

In 2005, a seminal joint report by the US Department of Energy and the US Department of Agriculture tried to answer this question. Entitled "Biomass as a Feedstock for a Bioenergy and Bioproducts Industry: The Technical Feasibility of a Billion Ton Annual Supply,"[11] and commonly called the Billion Ton Study, it provided a scenario that identified 1366 million dry tons of forest and agricultural biomass that could be sustainably available for bioenergy. The scenario included estimates for increasing crop yields, improved harvesting methods, and current resource uses. Most importantly, only modest land use changes are assumed to maintain sustainable bioenergy production and avoid unintended adverse environmental impacts such as deforestation, reduced soil health, and displacement of food crops for energy production. A shift of $<10\%$ of available cropland, cropland pasture, and idle land could account for almost half of the agricultural biomass required for bioenergy production.

The Billion Ton Study addressed the question of resource availability with the potential to reduce petroleum consumption for transportation by 30%; however, it lacked cost information. In 2011, an update was published, "U.S. Billion Ton Update: Supply for a Bioenergy and Bioproducts Industry," that included supply curves for identified resources with cost information and a more granular spatial distribution of feedstocks.[12] However, costs for agricultural materials accounted for only preparation and delivery to the farmgate and costs for forest materials only accounted for accumulation at the roadside. The latest update, in 2016,[13] is in two volumes. The first is on the economic viability of feedstocks delivered to the biorefinery and includes the cost of additional preparation and transportation and how that might affect feedstock availability. The second summarizes the environmental effects of specific scenarios to provide a more critical assessment of sustainability.

Biomass Conversion

With enough biomass available, technology development was focused on efficiently and economically converting these sustainable resources to displace a large fraction of fossil fuel, which would have a positive impact on the future environmental intensity of energy consumption. Adapting known technology is a strategy for accelerating development and leveraging existing infrastructure to help manage costs and support deployment.

The first stage is transporting the biomass from the source to the biorefinery. Supply chain options from the agricultural and forest products industries can be adapted to optimize harvesting, preparation, and transportation. Once inside the plant gate, it is converted into intermediates that are upgraded into biofuels, bioproducts, or biopower. The conversion process is either biochemical or thermochemical, as shown in Fig. 1.6. Chapter 3 gives a detailed description of each.

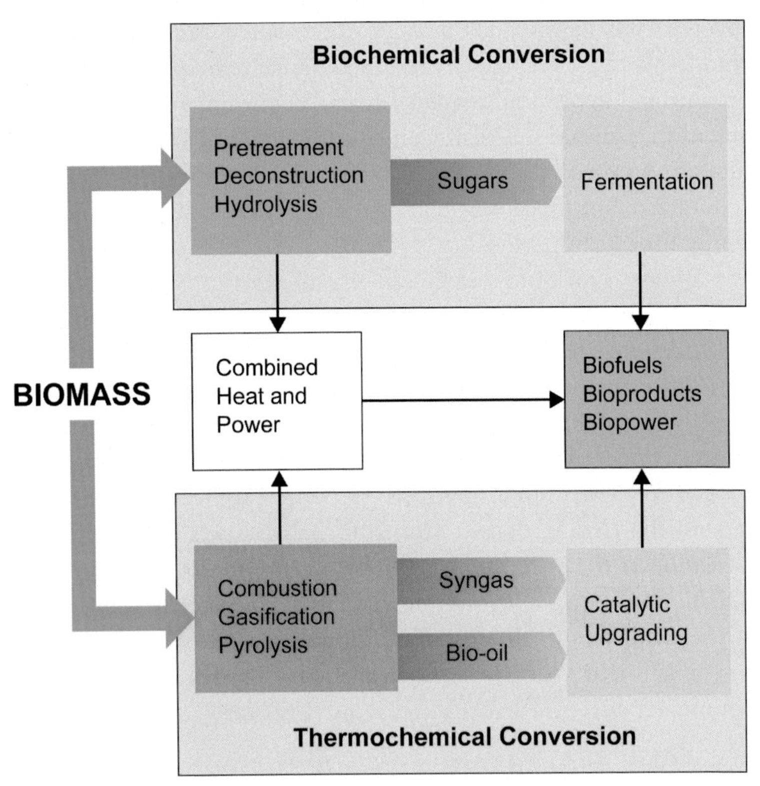

Fig. 1.6 Simplified description of biomass conversion.

Biochemical conversion starts by breaking down the cellulose and hemicellulose components into sugars that can be fermented to produce ethanol or other biochemicals. Biomass deconstruction is the core of the pulp and paper industry, where chemicals are used to dissolve lignin, and hemicellulose and cellulose fibers are recovered. Fermentation is the basis of the brewing and distilling industries. The general concepts for biochemical conversion to biofuels and bio-chemicals are well known from existing industries, but development is still required to optimize technology for biomass applications. Two technical challenges remaining for biochemical conversion process development are low cost sugar recovery which is significantly impacted by enzyme cost and mitigation of inhibitory byproducts and impurities. An economic challenge is finding high-value uses for lignin instead of burning it for heat and power.

Thermochemical conversion uses heat to depolymerize biomass into gas phase or liquid phase intermediates that are catalytically upgraded into

Table 1.1 Physical and Chemical Properties of Biological and Fossil Resources.

	Biomass (wood)	Coal (bituminous)	Bio-oil	Petroleum crude
Bulk density (kg/L)	0.25–0.4	0.6–1.0	1.1–1.2	0.7–1.0
Energy density (MJ/kg)	18–23	24–35	18	42
		Physical properties (wt%, dry)		
Volatile matter	80	15–55		
Fixed carbon	15–20	45–85		
Ash	1–3	3–12		
		Elemental composition (wt%, dry)		
C	50	76–90	56	84
H	6	4	6	14
O	40	10–20	38	1
N	<1	1–2	<1	1
S	<0.05	0.7–4.0	<0.05	1–3

biofuels and bio-products. Thermochemical technologies are an integral part of fossil fuel conversion, and many have been adapted for biomass. Biomass gasification to produce syngas ($CO + H_2$) leverages reactor designs and gas cleanup and conversion methods developed for coal gasification. Similarly, biomass pyrolysis produces a liquid bio-oil intermediate that many have attempted to upgrade into transportation fuel using methods developed for petroleum refining. Adapting commercially available fossil fuel technologies for biomass thermochemical conversion has been challenging because biomass and bio-oils have substantially different physical and chemical properties compared to coal and petroleum, as shown in Table 1.1. Among the technical problems are improving carbon efficiency (yield) and quality (minimizing impurities and undesired chemical components) of thermochemical intermediates, syngas, and bio-oils. An economic challenge is minimizing capital cost at suitable scale.

Motivation

Since the RFS was enacted in 2005, the hydrocarbon landscape has changed dramatically. In 2008, oil prices spiked and ethanol became a lower cost option. Shortly after that, the economic downturn known as the Great Recession quickly tempered global demand for energy, and petroleum prices dropped. Increasing capacity in the ethanol industry and reduced gasoline consumption began to restrict the market for ethanol as a transportation fuel by what has become known as the 10% "blend wall."

In parallel, a new energy revolution was emerging in the United States, due in large part to advances in horizontal drilling and hydraulic fracturing to extract untapped oil and gas from shale formations. Shale oil production has caused deep cuts in fossil fuel imports, and the United States is well on the way to sourcing only North American reserves to satisfy domestic demand.[14] Shale production has also provided a sustained, low cost source of natural gas.[15]

The domestic energy landscape has changed in unexpected ways compared with the projections made when the RFS was devised. Gasoline consumption has dropped by 16% over the then-projected numbers,[1] and aggressive CAFE targets that are being proposed will accelerate that trend. Plug-in hybrid and all-electric vehicles are now commercially available, and future improvements in battery storage technology are expected. New oil and gas reserves are being tapped with innovative technology that has transformed the United States into a net energy exporter and has led the nation to energy independence.

Close-up: Impact of Electric Vehicles on Biofuels

In theory complete fossil fuel displacement for land transportation may be achieved with electric vehicles but in practice there may be some barriers to achieving this. Car makers are finally getting serious about traversing the main hurdle: battery cost. Several models are commercially available, and the market share for electric vehicles will likely increase in the short term because of government incentives and in the long term with future improvements in battery storage technology. When the Nissan Leaf and the Chevy Volt hybrid were first sold, lithium cells cost over $450 per kWh (kilowatt hour). Typically, each mile driven uses 0.25 kWh, so a hundred mile range will require 25 kWh in principle. But it is impractical to drain a battery down to zero charge, so a 100 mile range needs a battery pack with about 30 kWh, 15–20% excess charge capacity.

For serious market penetration of electric vehicles, many have posited that battery costs had to drop below $200 per kWh, preferably to $150. At $150, a 30 kWh battery would cost about $5500, accounting also for the ancillary costs for the pack beyond that of the cells. That is a reasonable fraction of a vehicle selling price of $25,000, a reasonable target for a five passenger economy car. An all-electric car has no internal combustion engine, no transmission, and possibly no differential (if four motors are used), all of which reduces cost. But a 100 mile range may not sell broadly, and increasing range translates into higher battery costs. For a 200 mile range, the battery pack would cost $11,000, using the same $150 per kWh. This might drive the cost of the same five passenger economy car to a price point that the market will not bear.

Close-up: Impact of Electric Vehicles on Biofuels—cont'd

The market is different for luxury vehicles, where the battery cost is a lower percentage of the total vehicle cost. The cost differential between an all-electric luxury vehicle and a traditional gasoline powered luxury vehicle is not very significant, so a zero tailpipe emissions vehicle is an easy choice.

A hybrid vehicle has many of the positive features of an electric vehicle for driving conditions where internal combustion engines are less efficient, such as when the vehicle is not moving and for starting and low speed driving. Collectively, these features combined will typically add 40% or so to the gas mileage in the frequent start and stop of short range city driving as the range of the battery is extended by regenerative braking. Fewer benefits will be observed for long distance trips, but longer range is guaranteed with the gasoline engine available.

A hybrid five seat vehicle can deliver 45 mpg in the city. An all-electric will give about 105–110 mpg (computed on the basis of a gallon of gasoline containing 34 kWh of energy). It will cost more to purchase an all-electric, but maintenance should be much less. If gasoline prices remain low, fuel efficiency will have less impact on some vehicle purchase decisions until there is an economic or social benefit for the environmental emissions reduction. Cost-effective biofuels can bridge the gap by providing lower vehicle emissions as the propulsion system used in the transportation fleet evolves.

Vikram Rao

Research Triangle Energy Consortium, Research Triangle Park, NC, United States

References

1. U.S. Energy Information Administration. *Annual Energy Outlook 2015 With Projections to 2040*. April 2015. DOE/EIA-0383(2015).
2. Cai X, Stiegert KW. Market analysis of ethanol capacity. *Int Food Agribus Manag Rev*. 2014;17(1):83–93.
3. Renewable Fuels Association. 2015 Ethanol Industry Outlook. www.ethanaolRFA.org; 2015.
4. *Reducing the Fuel Consumption and Greenhouse Gas Emissions of Medium- and Heavy-Duty Vehicles, Phase Two: First Report*. Washington, DC: The National Academies Press; 2014.
5. Galadima A, Muraza O. From synthesis gas production to methanol synthesis and potential upgrade to gasoline range hydrocarbons: a review. *J Nat Gas Sci Eng*. 2015;25:303–316.
6. Camobreco V. *Inventory of U.S. Greenhouse Gas Emissions and Sinks 1990–2017*. Washington, DC: United States Envrionmental Protestion Agency; 2019. April 11, 2019. EPA 430-R-19-001.
7. Stock JH. *The Renewable Fuel Standard: A Path Forward*. Columbia Center on Global Energy Policy; 2015.

8. Mauser W, Klepper G, Zabel F, et al. Global biomass production potentials exceed expected future demand without the need for cropland expansion. *Nat Commun*. 2015;6:8946.
9. Henry RC, Engström K, Olin S, Alexander P, Arneth A, Rounsevell MDA. Food supply and bioenergy production within the global cropland planetary boundary. *PLoS ONE*. 2018;13(3). e0194695.
10. IRENA. *Global Energy Transformation: A Roadmap to 2050 (2019 edition)*. Abu Dhab: International Renewable Energy Agency; 2019. April 2019.
11. Perlack RD, Wright LL, Turhollow AF, Graham RL, Stokes BJ, Erbach DC. *Biomass as a Feedstock for a Bioenergy and Bioproducts Industry: The Technical Feasibility of a Billion-Ton Annual Supply*. April 2005. DOE/GO-102005-2135.
12. Turhollow A, Perlack R, Eaton L, et al. The updated billion-ton resource assessment. *Biomass Bioenergy*. 2014;70:149–164.
13. Langholtz MH, Stokes BJ, Eaton LM. *2016 Billion-Ton Report: Advancing Domestic Resources for a Thriving Bioeconomy*. EERE Publication and Product Library; 2016 DOE/EE-1440; Other: 7439 United States https://doi.org/10.2172/1271651 Other: 7439 EE-LIBRARY English.
14. Chew KJ. The future of oil: unconventional fossil fuels. *Philos Trans R Soc A Math Phys Eng Sci*. 2014;372(2006):20120324. https://royalsocietypublishing.org/doi/10.1098/rsta.2012.0324.
15. Wang Q, Chen X, Jha AN, Rogers H. Natural gas from shale formation—the evolution, evidences and challenges of shale gas revolution in United States. *Renew Sust Energ Rev*. 2014;30:1–28.

Additional Reading

Fouquet R. The slow search for solutions: lessons from historical energy transitions by sector and service. *Energy Policy*. 2010;38(11):6586–6596.

CHAPTER TWO

Biomass Characterization

Introduction

What is biomass? The answer depends on who is asking the question. For example, there were 14 definitions of biomass in United States legislation and the US tax code between 2004 and 2010.[1] Here are two basic definitions:

- From an ecological perspective, the amount of living matter in a given habitat, expressed either as the weight of organisms per unit area or as the volume of organisms per unit volume of habitat.
- From an energy perspective, organic matter, especially plant matter, that can be converted to fuel and is therefore regarded as a potential energy source.

If we consider biomass as an energy source, then it can be defined as organic matter that captures sunlight to convert CO_2 and water, through photosynthesis, into stored chemical energy (carbohydrates) for growth. The CO_2 that is converted to fixed carbon in biomass is indiscriminate of renewable or fossil-derived sources, but biomass is considered a renewable resource because it is grown, used, and regrown on a comparatively short time scale in large quantities.

From a global perspective, there are two broad types of biomass: terrestrial and aquatic. Since 70% of the Earth's surface is covered by water, one might assume that aquatic biomass is more abundant. In fact, there are roughly equal amounts of terrestrial and aquatic biomass available, but terrestrial concentrates more carbon per unit area. The aquatic type consists of microalgae (phytoplankton) and macroalgae (kelp and other plants). The terrestrial type is distinguished as woody (hardwoods and softwoods) and herbaceous (grasses and agricultural crops). Food, feed, fiber, and forest products are the primary uses for most terrestrial biomass resources because they yield higher value than energy production. Grains (corn, wheat, rice, barley, etc.) and oilseeds (soybeans, canola, peanuts, etc.) are food crops for human and animal consumption. Grasses (switchgrass, miscanthus,

Analytical Methods for Biomass Characterization and Conversion
https://doi.org/10.1016/B978-0-12-815605-6.00002-0

Fig. 2.1 Miscanthus growing in a field. *(Photo by Dennis Schroeder, NREL 47108.)*

Bermuda, rye, alfalfa, etc.) and meals and residues from food crop production are used as animal feed. Fig. 2.1 shows a field of mature miscanthus ready for harvest.

Residues from these primary applications are also biomass resources and are used for bioenergy production. Woody biomass residues include thinnings from forestry operations, sawmill residues from lumber production, and bark and black liquor from pulp and paper production. Agricultural residues such as corn stover (Fig. 2.2) and wheat straw are remnants of grain harvesting that can be used for animal feed or bioenergy.

Fig. 2.2 Corn stover, the residue remaining after corn is harvested. *(Photo by David and Associate, NREL 02897.)*

Other biomass-derived waste streams are available for bioenergy. These include municipal solid waste that contains paper, cardboard, and discarded food; construction and demolition wastes (wood); animal manures; wastewater treatment sludge (bio-solids); and biogas (landfill and anaerobic digestion gases rich in methane).

With such a wide variety, the answer to "What is biomass?" depends on the end use. The level of detail for characterizing resources varies from a basic measurement of how much there is in a given volume to a comprehensive chemical analysis of the major and minor organic and inorganic constituents in a given feedstock. Regardless of the level of detail, biomass characterization is needed to understand what a resource is and, more importantly, how to maximize the utilization efficiency. The simplest bioenergy application is biomass combustion to recover the stored energy. For combustion, energy content is the most important property. Biofuels applications require a much more detailed chemical characterization to optimize the thermal, chemical, or biological deconstruction to produce intermediates that are converted into biofuels, bioproducts, and biopower. Detailed chemical characterization becomes critical for developing new technologies. Comprehensive analyses are needed to understand new chemistries and processes and to optimize process variables and materials to maximize carbon conversion and energy efficiency. The following sections highlight characterization needs and methods for specific bioenergy applications.

Physical Properties

Basic physical property measurements of biomass resources are useful for defining preparation and handling needs. Size reduction is often required before biomass is introduced into a conversion process. Selection of a method for size reduction (chipping, grinding, or chopping) depends on the type of feedstock. Woody material is often chipped to 5 cm or less; grinding reduces the size to 2 mm or less. Chopping is required for herbaceous material. Two important considerations when measuring particle size are that (1) size reduction produces a distribution of sizes, not a single uniform one, and (2) particles are usually not spherical but have aspect ratios of 2 or greater. Particle size distribution can be measured optically by laser scattering with a particle size analyzer or by using a set of screens with carefully calibrated mesh size to sieve the biomass sample. Particle size distribution is then determined by weighing the amount of material that passes through each of the screens. If the aspect ratio is high, as with some grasses, straws, and agriculture residues, the distribution may be skewed. Size reduction also affects the bulk density of the feedstock. Size distribution, aspect ratio, and bulk density are all important for designing robust and reliable biomass feeding systems.

Most processes are scaled based on the available dry matter, so moisture content is a fundamental property that needs to be measured. It is most simply determined by placing a sample of a known weight in an oven at 105 °C and waiting until the sample weight no longer changes. The loss in weight is a measure of the moisture content. It is important to maintain the sample temperature at 105 °C to avoid loss of any volatile matter and to prevent ignition.

Moisture content is one of the measurements done in a proximate analysis, which also determines volatile matter, fixed carbon, and ash content of a sample following ASTM D5142. A typical proximate analysis is conducted by loading a small sample (~10 mg) into a thermogravimetric analyzer (TGA). The temperature is increased at a rate of 10 °C/min in helium (or other inert gas) to 105 °C. The weight loss is a measure of moisture content. The change in weight as the temperature is increased from 105 °C to 600 °C at a rate of 10 °C/min is a measure of the volatiles content. At 600 °C, the gas is switched from inert gas to air. The change in weight is a measure of the fixed carbon in the sample that combusts. The residual mass left after oxidation is the amount of ash in the sample. A typical TGA profile for the proximate analysis of a solid sample is shown in Fig. 2.3.

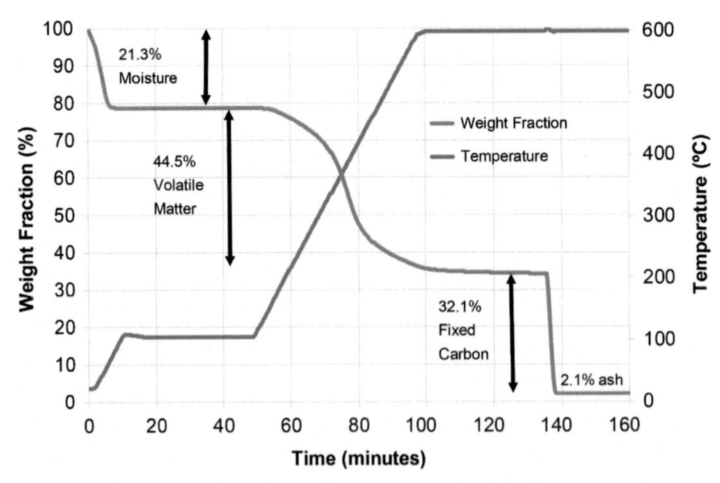

Fig. 2.3 Typical thermogravimetric analyzer pattern (weight fraction and temperature) for proximate analysis of a solid sample.

Elemental Composition

Ultimate analysis is a measure of the elemental composition (CHONS) of biomass and is done according to ASTM D5373. Samples are oxidized at high temperature (~1000 °C) in a furnace, and the combustion products (CO_2, H_2O, N_2, and SO_2) are separated on a column and analyzed by a thermal conductivity detector (TCD). In the case of oxygen determination, the samples are pyrolyzed in the furnace to form N_2, CO, and H_2. The CO is then separated from other gases, and the eluted gases are analyzed by the TCD. However, the oxygen content is typically calculated by difference. Table 2.1 gives examples of proximate, ultimate, and

Table 2.1 Proximate, Ultimate, and Elemental Ash Analyses for Selected Biomass Reference Materials Characterized at Idaho National Laboratory (https://bioenergylibrary.inl.gov/Sample/BiomassInfo.aspx).

	Lodgepole pine	Hybrid poplar	Switchgrass	Miscanthus	Wheat straw	Corn stover
Proximate analysis (wt%, dry basis)						
Volatile	84.5	86.48	80.2	85.53	77.04	79.0
Fixed carbon	14.41	12.65	15.6	13.06	13.89	16.7
Ash	1.08	0.87	4.2	1.4	9.07	4.3
Ultimate (wt%, dry basis)						
C	50.14	49.40	47.20	50.64	45.02	48.70
H	3.06	6.03	5.70	5.85	5.90	5.70
O[a]	46.80	44.57	47.05	41.88	38.82	44.90
N	–	–	0.05	0.21	1.06	0.70
S	–	–	–	0.01	0.12	–
Ash composition (wt% metal in ash reported as oxide)						
Al_2O_3	5.19	1.02	0.25	0.29	2.77	0.28
CaO	10.98	27.02	7.37	18.34	10.83	8.99
Fe_2O_3	6.87	0.71	1.63	1.20	2.99	1.12
K_2O	7.63	25.56	17.55	6.44	15.45	26.38
MgO	3.24	7.00	9.79	9.03	2.69	6.09
MnO	0.53	0.07	0.19	1.11	0.07	0.09
Na_2O	6.94	1.41	1.61	0.18	1.16	0.08
P_2O_5	1.57	5.97	4.45	3.58	2.15	2.79
SiO_2	44.26	10.81	53.53	52.31	58.16	51.99
TiO_2	0.23	0.07	0.01	0.02	0.11	0.01
SO_3	3.19	2.98	2.73	3.15	2.34	2.20

[a]Oxygen reported by difference.

elemental ash analyses for several biomass feedstocks. These feedstocks are standard reference materials maintained in the Idaho National Laboratory's Biomass Feedstock National User Facility as part of their Bioenergy Feedstock Library. The literature contains many more analyses of a wide variety of biomass resources, but these reference materials provide an illustrative example that biomass is mostly carbon, hydrogen, and oxygen, and the relative composition of these major components does not vary much. In fact, the chemical formula of biomass is often represented as $CH_{1.8}O_{0.5}N_{0.2}$. Nitrogen and sulfur are minor components, and their relative concentrations are a function of biomass type and growing conditions.

Elemental analysis is important for determining element balances in conversion processes, particularly carbon balances for carbon efficiency calculations. Biomass energy content can be measured directly by bomb calorimetry; however, correlations based on proximate and ultimate analyses have been developed with quite good accuracy. The following relationship can be used to calculate the higher heating value (HHV) of a sample based on the proximate and ultimate analyses[2]:

$$HHV = 3.55C^2 + 232C - 2230H + 51.2(HC) + 131N + 20,600 \quad (2.1)$$

where C, H, and N are mass percent of carbon, hydrogen, and nitrogen, respectively. Energy content expressed as HHV assumes that any water produced by combustion is liquid. Subtracting out the energy required to vaporize water during sample combustion leads to energy content expressed on a lower heating value basis.

In bioenergy applications, sulfur and nitrogen emissions from combustion can have adverse environmental implications. Some feedstocks have high concentrations of chlorine, and combustion can cause corrosion in high temperature systems from HCl formation, particularly as the flue gas cools and water is condensed. Another concern with high chlorine feedstocks is dioxin emissions in combustion systems. Sulfur, nitrogen, and chlorine are also considered catalyst poisons that could negatively affect the performance of biomass conversion processes.

The unconvertable, refractory residue known as ash is segregated into two fractions. Certain biomass materials assimilate inorganic components for biological processes and structural integrity. Also, ash can contain mineral matter that comes from soil collected with the feedstock. Regardless of the origin, the amount and composition of ash has a significant impact on bioenergy and biofuels applications. At a minimum, the ash has no energy value and cannot be converted or upgraded into fuels. Therefore, the higher the ash content of a feedstock, the lower the energy efficiency of a given process.

Analytic methods for determining the elemental composition of coal ash are followed for biomass ash. ASTM D3174 is used to determine the residue from burning coal and coke, that is, the ash content of a solid material as reported in the proximate and ultimate analyses. ASTM D3682 gives standard test methods for determining the compositional analysis of ash residues from solid fuels. The ash preparation method (combustion at 600 °C in excess air) can alter the chemistry of the mineral matter. Metal carbonates can decompose and release CO_2, pyrites can oxidize and release SO_2, and some alkali metals are volatile at preparation temperatures. Therefore, ash components are usually stated as wt% of the metal in the ash reported as an oxide. ASTM D6349 extends these analyses to ash byproducts that accumulate in combustion systems, such as fly ash, bottom ash, slag, and clinkers. These samples are digested in acid and analyzed by inductively coupled plasma atomic emission spectroscopy.

Most woody biomass has low ash content, typically 1 wt% or less. Herbaceous biomass and agricultural residues have significantly higher ash. The corn stover and wheat straw reference materials contain 4 wt% and 9 wt% ash, respectively. Some rice straws and rice hulls contain up to 20 wt%. Regardless, most of the ash tends to be silica, as shown in Table 2.1. Alkaline earth metals and alkali metals are also major components and vary as a function of the type of feedstock.

Biomass combustion and biomass co-firing with coal underscore the impact of high temperature ash chemistry as it relates to fouling and slagging in commercial scale power systems.[3] Feedstocks can have high concentrations of potassium in ash that is volatile at combustion temperatures. Alkali metal vapors, in conjunction with silica and alumina in ash, can form low melting point eutectic mixtures that deposit on surfaces. At the right temperature, these surfaces become sticky and promote deposit formation as particulate matter adheres and accumulates.

Chemical Composition

The ability to understand and measure the chemical composition of biomass feedstocks defines the input for developing bioenergy processes. It provides the basis for determining process performance in terms of carbon conversion efficiency and energy efficiency and facilitates process optimization to maximize product yields. Wet chemistry methods have been developed for this measurement. These multi-step methods are labor intensive but accurate and reproducible. Spectroscopic techniques are also being developed that provide lower-cost, rapid analyses that are better suited for process development.

The chemical composition of a feedstock is a function of plant species, plant genetics, growing environment, harvesting and collection methods, and storage duration and conditions. These compositional variations can be measured and studied, but biomass is inherently inhomogeneous. Within a given tree, plant, field, etc., composition depends on where the sample is collected. Imagine taking a 100-g sample for analysis from a process that uses 2000 tons per day. Getting a representative sample can take a lot of time and effort that is often not justified in a given process environment. Nevertheless, chemical composition is a variable that must be understood and controlled to determine impact and sensitivity in process development.

The chemical structure of biomass is a complex, integrated network of three structural biopolymers—cellulose, hemicellulose, and lignin—with other organic constituents called extractives that depend on the type of material. Cellulose is the most abundant organic component, ranging from 38% to 50% by weight in woody and herbaceous feedstocks. It is a polysaccharide that consists of repeating hexose (glucose) units. Both crystalline and amorphous cellulose are present. Crystallinity is defined by the degree of polymerization, that is, a measure of the chain length. The relative fraction of crystalline and amorphous becomes important for developing pretreatment processes to deconstruct biomass for biofuels production as described in Chapter 3. Hemicellulose is the second most abundant at 23–32%. It has a less ordered structure than cellulose and is a polysaccharide composed of pentoses. Xylan is the most common pentose in hemicellulose. The third most abundant is lignin, a highly branched biopolymer with methoxylated aromatic subunits, at 15–25% by weight. Basic lignin monomeric structures, shown in Fig. 2.4, are *p*-coumaryl alcohol, coniferyl alcohol, and sinapyl alcohol, which combine to form guaiacyl (G), syringyl (S),

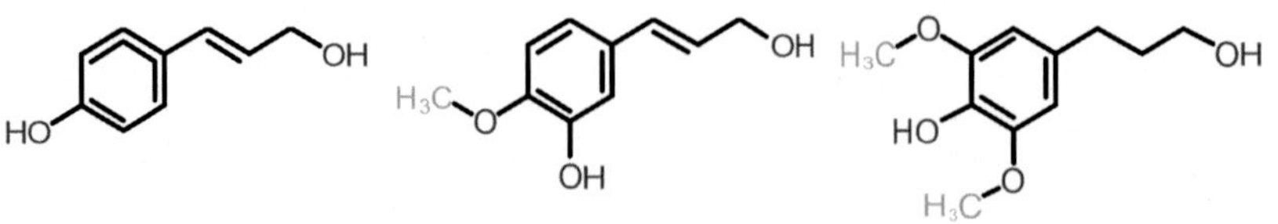

p-Coumaryl Alcohol **Coniferyl Alcohol** **Sinapyl Alcohol**

Fig. 2.4 Monolignol structures.

and *p*-hydroxyphenyl (H) units in the lignin. The exact structure of lignin is difficult to characterize and varies widely for different feedstocks. Therefore, lignin is often defined by the relative number of syringyl groups compared to the number of guaiacyl groups, the S/G ratio.

Extractives is the general term for non-structural organic components. They vary quite substantially for different materials and include gums (turpentines and rosin), resins, fatty acids, waxes, sterols, proteins, tannins, terpenes, and lipids. Softwoods such as pine have high concentrations of terpenes and rosins. Legumes have high concentrations of proteins. Oil seeds and algae have high concentrations of lipids.

Wet Chemical Methods

Measuring the chemical composition of biomass requires first breaking it down into individual components, then analyzing each component separately, and finally combining the measurements together with 100% mass closure to make sure nothing was missed or double counted. The type of biomass dictates which methods should be used, and the order in which the procedures are performed is important for quality assurance, reproducibility, and accuracy. A comprehensive set of Laboratory Analytical Procedures for biomass compositional analysis have been developed at the National Renewable Energy Laboratory.[4]

The first step is preparing the sample. Preparation includes drying it to 10 wt% moisture and reducing the particle size to 2 mm or less. The sample is air dried or oven dried, depending on the climate, until the change in weight is less than 0.1% per hour of drying time. Accurately determining moisture content is important because the analysis results are reported on a dry basis.

Non-structural extractives are removed first by using water, solvents, or a combination of both. Herbaceous feedstocks can contain a significant fraction of water-extractable material that includes protein, some ash components, and some sugars (like sucrose). Ethanol extraction is used for all biomass feedstocks to remove waxes, gums, terpenes, and fatty acids. The large number of compounds in low concentration in the extracted liquids precludes the quantitative identification of individual species, so extractives are generally treated as a whole.

Ash content of the starting material is a measure of the inorganic content of the extracted material and the inorganic content of the extracted ash. The extractable ash is determined by the difference in the ash content of the starting material and that of the extracted solid material. Both samples are oxidized in a furnace at 575 °C to remove all organic matter. The remaining residue is the ash. A similar approach is taken for protein content. The nitrogen content of the starting biomass and the extracted material is determined separately. The difference can be correlated to the protein content of the sample.

The extracted material contains the remaining cellulose and lignin. A two-step analytical hydrolysis procedure is used to separate the cellulose and lignin and chemically decompose the cellulose into its constituent monomeric sugars. The hydrolysate is neutralized, and the sugars are analyzed and quantified by high performance liquid chromatography. The lignin is segregated into two fractions, acid-insoluble and acid-soluble. The acid-insoluble material is high molecular weight lignin. The weight of the acid-insoluble lignin is corrected for ash and protein content. The acid-soluble lignin is analyzed and quantified by ultraviolet-visible spectroscopy.

As an example, the wet chemical analyses of the reference materials from Idaho National Laboratory's Biomass Feedstock Library are presented in Table 2.2. The analytical methods followed the Summative Mass Closure Laboratory Analytical Procedure developed at NREL (NREL/TP-510-48087). Of note is that no protein was measured for the woody feedstocks, but the pine and poplar had the highest lignin content and relatively low hemicellulose (xylan) content. The herbaceous feedstocks had less lignin, measurable protein, and more hemicellulose.

Spectroscopic Methods

The wet chemical methods for determining composition are reliable and reproducible but require substantial time and effort. Consequently, rapid analysis methods are sought to reduce cost and increase throughput to support process development. Fourier transform infrared (FTIR) and near infrared (NIR) spectroscopy have been used, but the availability of robust, cost-effective instruments that have been proven in industrial settings has propelled NIR as the rapid analysis method of choice.[5]

The success of NIR is a function of the robustness of the calibration model built from the compositions of standard materials determined by wet chemical methods. The accuracy and precision of the wet methods used for calibration is retained in the rapid analysis, so there is no tradeoff between speed and accuracy and precision. NIR reflectance spectra are collected for any number of samples and correlated with the calibration

Table 2.2 Chemical Compositions of Standard Reference Materials From Idaho National Laboratory's Biomass Feedstock Library (https://bioenergylibrary.inl.gov/Sample/BiomassInfo.aspx).

Weight %	Lodgepole pine	Hybrid poplar	Switchgrass	Miscanthus	Wheat straw	Corn stover
Structural ash	0.27	0.24	1.88	0.52	5.50	2.37
Extractable inorganics	0.13	0.50	2.07	0.02	3.37	1.48
Structural protein	–	–	1.51	0.47	3.07	1.84
Extractable protein	–	–	0.54	0.16	1.19	0.72
Water extracted glucan	0.13	0.45	2.28	0.27	1.56	0.62
Water extracted xylan	0.04	0.05	0.09	0.10	0.92	0.47
Water extractives other	2.45	1.94	6.68	2.61	4.76	4.46
Ethanol extractives	1.74	2.04	2.68	2.05	2.76	2.73
Lignin	30.5	25.70	16.24	20.41	16.27	16.37
Glucan	41.37	43.78	33.21	40.79	32.24	35.45
Xylan	5.9	13.29	21.65	22.01	16.95	22.34
Galactan	2.84	1.42	1.43	1.11	1.60	1.54
Arabanan	12.3	2.76	3.27	2.86	3.17	3.40
Acetate	1.17	4.24	3.07	3.87	1.70	1.58
Total	98.84	96.41	96.60	97.25	95.06	95.37

model. Multivariate statistical analysis is used to regress the spectroscopic data against the calibration data to translate the NIR fingerprints into biomass compositions. Compositions can be determined in real time, so conversion process conditions can be adjusted to account for variations in feedstocks. This provides real-time feedback to support process development and optimization.

The heart of NIR rapid analysis is the calibration models that are typically built for selected varieties of a given feedstock. For example, a model for corn stover can be developed from selected cultivars and varieties that are commonly grown commercially but have enough variance in chemical composition to cover a broad variable space. Unfortunately, the variation in chemical composition for different types of biomass is too large to develop a robust universal calibration model. Consequently, separate models are needed for different materials. NIR is fast and robust once the calibration models are developed and validated, but effort is required to develop specific calibration models for specific feedstocks.

Analytical Pyrolysis

Pyrolysis for biofuel production is discussed in detail in Chapter 3, but pyrolysis can also be used to understand biomass chemical composition. In contrast to the wet methods that chemically deconstruct the components, analytical pyrolysis relies on heat to thermally decompose biomass in an inert environment. Key challenges are (1) rapid heat transfer (both heating and quenching products) to minimize secondary reactions that can confound the results, and (2) comprehensive pyrolysis vapor analysis. The rapid heat transfer requirements can be met using small samples, on the order of 100–1000 μg. Of course, this requires precision measurements and challenges the detection limits for analyzing the pyrolysis products.

Micropyrolyzers with rapid heating elements for controlling the temperature of microreactors are integrated with a gas chromatography/mass spectrometry (GC/MS) instrument for detection of products (Fig. 2.5). Samples are placed is small cups or boats, the microreactor is purged, and the samples are dropped into the heated zone maintained at pyrolysis temperatures (400–600 °C). The sample is rapidly heated (~1000 °C/s) and the pyrolysis products are swept out of the reactor with an inert carrier gas, usually helium. There are two designs for commercially available micropyrolyzers. The first separates the reactor from the GC/MS and uses a heated capillary to deliver the pyrolysis products to the source oven in the GC/MS. The second

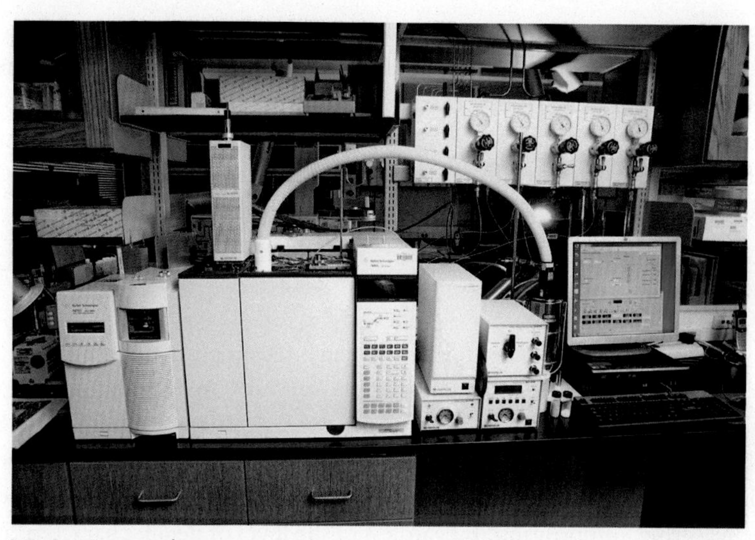

Fig. 2.5 A micropyrolyzer mounted on top of a GC/MS instrument. *(Photo by John Theilgard, RTI International.)*

mounts the reactor directly on the GC/MS such that the products are swept directly into the source oven.

The temperature limitations of the transfer capillary and the GC column itself may prevent some of the products from making it to the MS detector. Quantifying the components can help to determine what fraction of the sample is not detected. The data should be carefully evaluated to determine how comprehensive the sampling method is, and correlating the pyrolysis products to the actual biomass composition can be difficult. Specific thermal decomposition products can be associated with cellulose (levoglucosan and furan), hemicellulose (furfural), and lignin (guaiacol and syringol). The relative concentration of guaiacol and syringol is used to determine the S/G ratio and understand the structure of lignin in a given sample.

A unique method for analytical pyrolysis is molecular beam sampling mass spectrometry (MBMS),[6, 7] discussed in detail in Chapter 5. A high-temperature reactor is used to generate pyrolysis vapors that are sampled directly into the ionization region of a mass spectrometer. This has the advantage that all pyrolysis products can be sampled instantaneously without any loss due to the temperature limitations of capillaries and columns. Sampling all pyrolysis products instantaneously is also a disadvantage, as there is no chromatographic separation, making it difficult to identify and quantify individual sample components. Each mass spectrum is a complex linear combination of the signal intensities generated from hundreds of pyrolysis products. Multivariate statistical analysis is necessary for deconvoluting the data to identify specific components. The real utility of this method is determining the chemical variance between different feedstocks.

Conclusions

Characterizing biomass feedstocks can be as simple as defining the moisture content and bulk density of a sample and as complex as knowing the monomeric subunits that constitute the integrated network of the three structural biopolymers cellulose, hemicellulose, and lignin. The bioenergy application usually dictates the level of detail needed for understanding and measuring composition. Regardless of the level of detail, characterization supports the determination of mass and energy balances when coupled with a comprehensive analysis of various process streams.

Biomass is a heterogeneous material that requires diligent sample preparation and careful analysis to yield precise and accurate measurement of composition. The numerous standard analytical methods are summarized in Table 2.3. Compositional variance between different types of biomass can be quite large. Moisture is the component that can vary the most between feedstocks, ranging from percent solids in water to moisture content of solids. Moisture content has a direct impact on energy efficiency of biomass utilization. However, most compositions are reported on a dry basis to highlight the elemental and chemical variations between feedstocks.

Ash is the next most variable component in biomass. It has no energy or chemical value, so it has a direct impact on the efficiency of a given bioenergy application. Ash can be detrimental to a process by creating unwanted slag and deposits in combustion systems or acting as a catalyst poison or inhibitor in conversion processes. The organic content of biomass is divided into structural (cellulose, hemicellulose, and lignin) and non-structural (extractives) components.

Compositional analysis is the foundation for understanding biomass conversion chemistry and developing novel biofuel processes. Variations in organic content of different feedstocks can be quite large, but more subtle variations between different varieties of the same feedstock type can affect conversion process performance. Wet chemical methods have proven to

Table 2.3 Summary of Standard Methods for Biomass Compositional Analysis.

ASTM number	Title	Significance and use
ASTM E1721-01 (2015)	Standard Test Method for Determination of Acid-Insoluble Residue in Biomass	The acid-insoluble residue content is used in conjunction with other assays to determine the total composition of biomass samples.
ASTM E1758-01 (2015)	Standard Test Method for Determination of Carbohydrates in Biomass by High Performance Liquid Chromatography	The percentage, by mass, of sugar content is used in conjunction with other assays to determine the total composition of biomass samples.
ASTM E1690-08 (2016)	Standard Test Method for Determination of Ethanol Extractives in Biomass	Ethanol extractives are any materials found in biomass that are soluble in ethanol. They are not considered to be part of the structural components of biomass and should be removed prior to any chemical analysis of the sample. The prolonged extraction removes nonstructural materials that can include waxes, fats, resins, tannins, gums, sugars, starches, and pigments. Removing hydrophobic materials from the biomass makes it easier to wet the material for the analysis of structural components in the biomass.
ASTM E1756-08 (2015)	Standard Test Method for Determination of Total Solids in Biomass	Moisture is a ubiquitous and variable component of any biomass sample. It is not considered a structural component of biomass and can change with storage and handling of biomass samples. The determination of the total solids content allows for the correction of biomass samples to an oven-dried solids mass that is constant for a particular sample.
ASTM E1721-01 (2015)	Standard Test Method for Determination of Acid-Insoluble Residue in Biomass	

Continued

Table 2.3 Summary of Standard Methods for Biomass Compositional Analysis.—cont'd

ASTM number	Title	Significance and use
ASTM E1821-08 (2015)	Standard Test Method for Determination of Carbohydrates in Biomass by Gas Chromatography	
ASTM E1755-01 (2015)	Standard Test Method for Ash in Biomass	
ASTM E1757-01 (2015)	Standard Practice for Preparation of Biomass for Compositional Analysis	The acid-insoluble residue content is used in conjunction with other assays to determine the total composition of biomass samples.
ASTM E1358-97 (2013)	Standard Test Method for Determination of Moisture Content of Particulate Wood Fuels Using a Microwave Oven	The structural carbohydrate content is used in conjunction with other assays to determine the total composition of biomass samples.
ASTM E1534-93 (2013)	Standard Test Method for Determination of Ash Content of Particulate Wood Fuels	The ash content is an approximate measure of the mineral content and other inorganic matter in biomass.
ASTM E870-82 (2013)	Standard Test Methods for Analysis of Wood Fuels	These methods can be used for the proximate analysis, ultimate analysis, and the determination of the gross caloric value of wood fuels.
ASTM E871-82 (2013)	Standard Test Method for Moisture Analysis of Particulate Wood Fuels	This method can be used to determine the total weight basis moisture of any particulate wood fuel meeting the specified requirements.
ASTM E872-82 (2013)	Standard Test Method for Volatile Matter in the Analysis of Particulate Wood Fuels	This method can be used to determine the percentage of gaseous products, exclusive of moisture vapor, of any particulate wood fuel meeting the specified requirements.
ASTM E873-82 (2013)	Standard Test Method for Bulk Density of Densified Particulate Biomass Fuels	This method can be used to determine the bulk density (or bulk specific weight) of any densified particulate biomass fuel meeting the specified requirements.

be accurate and reproducible for measuring gross and subtle differences in composition but require substantial time and effort. Rapid analysis methods that track multiple variables simultaneously are desired for process monitoring, control, and optimization.

References

1. Riedy MJ, Stone TC. *Defining Biomass—A Comparison of Definitions in Legislation*. Mintz, Levin, Cohn, Ferris, Glovsky, and Popeo, P.C; 2010.
2. Friedl A, Padouvas E, Rotter H, Varmuza K. Prediction of heating values of biomass fuel from elemental composition. *Anal Chim Acta*. 2005;544(1):191–198.
3. Baxter LL, Miles TR, Miles Jr TR, et al. The behavior of inorganic material in biomass-fired power boilers: field and laboratory experiences. *Fuel Process Technol*. 1998;54(1–3): 47–78.
4. Sluiter JB, Ruiz RO, Scarlata CJ, Sluiter AD, Templeton DW. Compositional analysis of lignocellulosic feedstocks. 1. Review and description of methods. *J Agric Food Chem*. 2010;58(16):9043–9053.
5. Hames BR, Thomas SR, Sluiter AD, Roth CJ, Templeton DW. Rapid biomass analysis. *Appl Biochem Biotechnol*. 2003;105(1):5–16.
6. Evans RJ, Milne TA. Molecular characterization of the pyrolysis of biomass. *Energy Fuel*. 1987;1(2):123–137.
7. Evans RJ, Milne TA. Molecular characterization of the pyrolysis of biomass. 2. Applications. *Energy Fuel*. 1987;1(4):311–319.

Additional Reading

Baxter L, DeSollar R. *Applications of Advanced Technology to Ash-Related Problems in Boilers*. Springer US; 1996.
Klass DL. *Biomass for Renewable Energy, Fuels, and Chemicals*. Elsevier Science; 1998.

CHAPTER THREE

Biomass Conversion

Biofuels Overview

First-generation biofuels (summarized in Fig. 3.1) start with a variety of feedstocks to produce either a sugar or lipid intermediate that is converted to ethanol or biodiesel, respectively. Corn and sugarcane are the two most economically competitive feedstocks for producing ethanol for gasoline substitution. Syrup or molasses is processed from sugarcane, and the sugar (sucrose) is easily fermented to ethanol. The process is equally straightforward with sugar beets or sweet sorghum, although not as economically competitive. Using corn is a bit more complicated because the feedstock structure (starch) is more complex and must be broken down to simple sugars prior to fermentation. Starch is separated from corn in either a dry milling or wet milling process. Hydrolysis with amylase enzymes is then used to break down the starch (polysaccharide) into sugars (monosaccharides like glucose, disaccharides, and trisaccharides) that are easily fermented to ethanol. Sugarcane ethanol production, predominately practiced in Brazil, and corn starch ethanol, predominately practiced in the United States, are economically viable and account for the vast majority of worldwide ethanol production, as shown in Fig. 3.2.

For biodiesel production, lipids from oilseeds are very simply converted through the transesterification process.[1] Candidate sources are rapeseed (with a variant known as canola), palm oil, and castor oil, to name a few. Transesterification is a base-catalyzed batch process that reacts alcohol (methanol or ethanol) with triglycerides (fats or oils) to produce biodiesel (long-chain mono-alkyl esters) and crude glycerol. The crude glycerol usually contains unreacted alcohol, unused catalyst, and soaps. This simple method can be employed in small-scale operations with relatively low capital cost. Although the process is simple, the economics of biodiesel production are challenging. The high cost of virgin vegetable oil has forced the industry to rely on lower cost waste fats, oils, and grease that are only available in limited quantities. Glycerin is a potentially useful byproduct but unreliable

Analytical Methods for Biomass Characterization and Conversion
https://doi.org/10.1016/B978-0-12-815605-6.00003-2

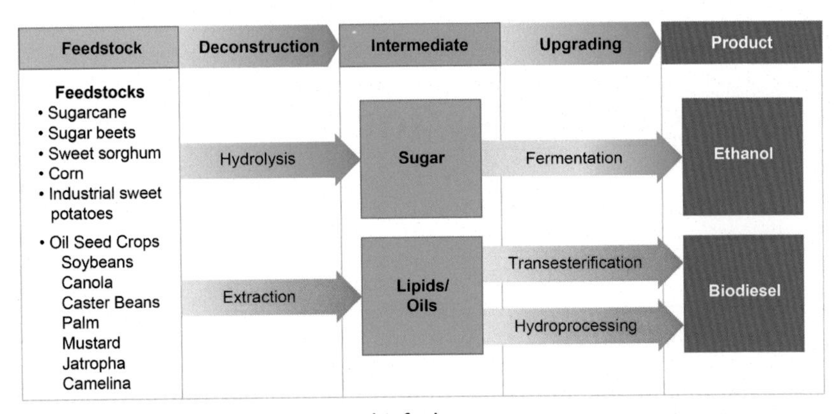

Fig. 3.1 Summary of first-generation biofuels processes.

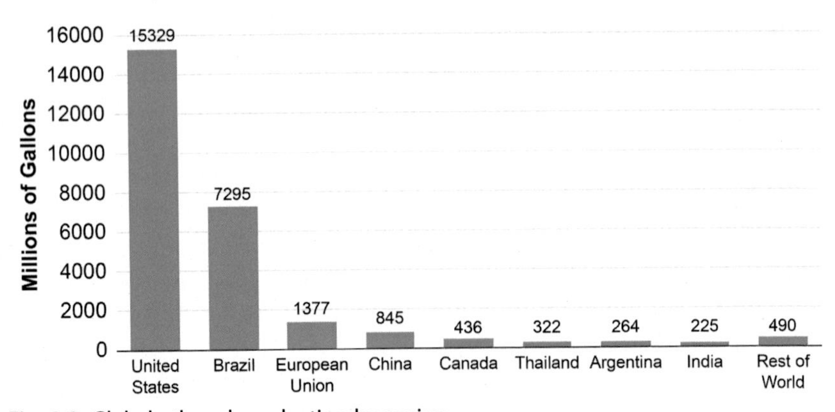

Fig. 3.2 Global ethanol production by region.

for improving biodiesel economics because of the high cost of purification and the price volatility of a relatively small market.

Alternative, nonfood oilseed crops are being developed for biodiesel production, such as jatropha. It is resistant to drought and pests and produces seeds containing 27–40% oil. A jatropha field will yield up to 1900 L of oil per hectare, compared with 1190 for canola and 446 for soybean.[2] Although there was early enthusiasm for jatropha, commercial application as a biofuel option will require additional efforts to domesticate promising varieties with optimal yields for a predictable supply.

Conventional hydroprocessing is being developed as an alternative to transesterification. Traditional refining technology has been adapted to convert triglycerides into biodiesel, commonly referred to as green or renewable diesel. The capital cost and complexity are greater than those associated with transesterification, but it is a continuous process that can be easily scaled up, and glycerin is not a byproduct.[3] Full-scale biorefineries have been built based on this technology. A primary challenge to commercial success is that the method is limited by the availability of low-cost feedstocks such as used cooking oils and greases and byproducts from the animal products industry (fats and tallows) or the wood products industry (tall oil).

A key technical challenge that has been overcome is the removal of phosphorous from the feedstock. Used cooking oils and other fats and greases contain varying amounts of phospholipids that can yield up to 1000 ppm of phosphorous in the feedstock. Phosphorous poisons hydroprocessing catalysts, so the phosphorous content must be reduced to <1 ppm for commercial success.

Biofuels from lignocellulosic biomass are commonly referred to as second-generation (cellulosic ethanol) and advanced (drop-in hydrocarbon replacements). While first generation technologies start with a modest separation process—starch hydrolysis and lipid extraction—second-generation and advanced technologies require more complicated and severe pretreatment steps to effectively deconstruct biomass to produce intermediate streams that can be upgraded into ethanol and hydrocarbon fuels.

Traditionally, second-generation and advanced biofuels technologies have been separated into biochemical and thermochemical conversion processes. These lines are getting blurred with the integration of the thermocatalytic process for upgrading biochemical intermediates and biological upgrading of thermochemical intermediates. Nevertheless, the basic lignocellulosic deconstruction of these intermediates involves either a chemical and enzymatic pretreatment to liberate a clean sugar intermediate that can be fermented to produce ethanol, or a thermal decomposition with or without an oxidizer to produce a gaseous (syngas) or liquid (bio-oil) intermediate. Fig. 3.3 summarizes these pathways.

Lately, there has been renewed interest in algae as an alternative feedstock. Algae are an excellent source of lipids that can leverage the technologies developed in the existing biodiesel industry for conversion into biofuels. The allure in part is that algae can be grown in ponds on non-arable land using brackish water. Algae growth can also be accelerated using carbon dioxide (CO_2) to enhance photosynthesis. In this way, algae is part of a climate change mitigation strategy when a source of fossil-derived CO_2 is used. But the scale required for cultivating and harvesting algae is daunting. Research on tailored species is ongoing. Also, there is at least one effort

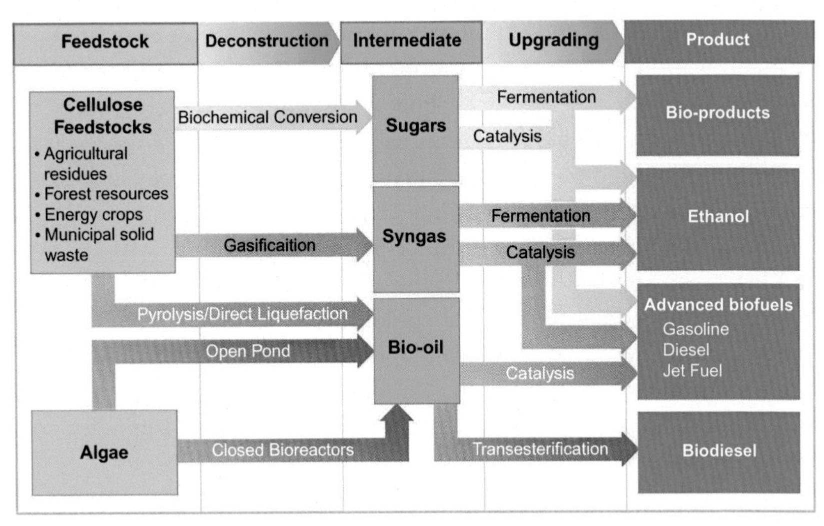

Fig. 3.3 Second-generation and advanced biofuels technology pathways using lignocellulosic and algal feedstocks.

underway to make algae production more of a factory operation, targeting lipid production while minimizing biomass growth.[4–6]

Biochemical Conversion

Biochemical conversion processes are like first-generation processes in that ethanol is often the desired product; however, the structural and biochemical differences between starch and biomass carbohydrates require more sophisticated methods to produce second-generation lignocellulosic ethanol. For corn ethanol, milling (dry or wet) is used to isolate the starch from the kernels; the starch is then hydrolyzed to produce monomeric sugars for fermentation to ethanol. The interwoven cellulose, hemicellulose, and lignin biopolymers in biomass present a complex recalcitrant structure that makes it more difficult and expensive to produce a clean cellulosic sugar stream that can be fermented to ethanol. Fig. 3.4 is a block flow diagram of a generic second-generation lignocellulosic ethanol process.

The first step in any second-generation process is the harvesting, delivery, storage, and comminution of biomass to prepare it for conversion. The similarity between the corn ethanol and lignocellulosic processes leads to a preference of agricultural residues as the feedstock. Cost can be minimized by collecting residues simultaneously as the grains (corn, wheat,

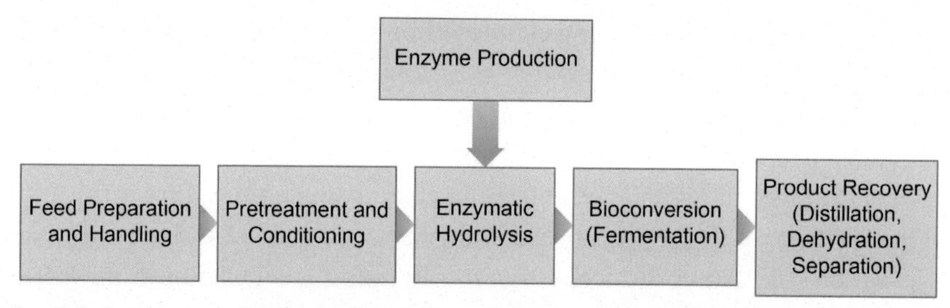

Fig. 3.4 A generic second-generation lignocellulosic ethanol process.

barley, etc.) are harvested. However, biochemical conversion processes are also being developed for woody biomass and grasses and are not limited to agricultural residues.

Biomass Pretreatment

The physical structure of lignocellulosic biomass requires a series of steps, called pretreatment, to deconstruct it and produce a clean sugar stream. Pretreatment is required to liberate the carbohydrate fraction from the structural components, mainly lignin, to expose cellulose and enhance enzymatic hydrolysis for producing cellulosic sugars. However, in nature, biomass varieties and phenotypes have genetically evolved to combat chemical, environmental, and biological degradation. Overcoming this recalcitrance is the primary challenge for pretreatment. Cellulose degree of polymerization (chain length and amorphous vs crystalline cellulose) and the extent to which lignin and hemicellulose bind with cellulose are contributors to recalcitrance. In general, pretreatment methods are designed to separate lignin and hemicellulose from cellulose, and the severity (temperature, residence time, and pH) of the process is related the recalcitrance of the feedstock. A variety of pretreatment methods are under development with the goal of maximizing sugar yield while minimizing degradation products that are fermentation inhibitors.[7–10]

Pretreatment processes are classified in several ways. At the highest level, they are separated as chemical or physicochemical. They are also distinguished by pH (acidic, neutral, and alkaline), the type of acid and other chemicals used, the temperature, and the pressure.

Chemical pretreatment of woody biomass has been practiced for over a century in the pulp and paper industry. In the kraft process, pulping chemicals (an aqueous mixture of sodium sulfide and sodium hydroxide) are mixed with softwood chips in digesters to remove hemicellulose and dissolve lignin while recovering cellulose fibers to produce wood pulp. Soda pulping is a sulfur-free process that uses sodium hydroxide as the primary delignification agent. Acidic pulping can also be done by the sulfite process that uses a mixture of sulfurous acid and either alkali hydroxides or alkaline earth carbonates to provide counter ions that produce the sulfite or bisulfite cooking liquor. Lignin is dissolved and can be recovered as lignosulfonates that can be sold as byproducts. Organosolv pulping is less common. It uses organic solvents with inorganic acid catalysts (hydrochloric acid or sulfuric acid). Common solvents are alcohols (methanol and ethanol), glycols, glycerol, and acetone.

Pretreatment methods have been developed based on these traditional pulping processes. Acid pretreatment incorporates concentrated or dilute inorganic acids such as HCl, H_2SO_4, H_3PO_4, and HNO_3. Acid pretreatment can be done before enzymatic hydrolysis or it can be used to simultaneously saccharify and hydrolyze cellulose and hemicellulose in one step. Concentrated acid pretreatment produces high sugar yields at relatively low temperature, but the process is expensive due to high acid consumption and energy demands for acid recovery. The dilute acid pretreatment requires higher temperature than concentrated acid processes to obtain comparable sugar yields and hydrolysis rates. Care must be taken in acid pretreatment (dilute or concentrated) to control process severity to avoid sugar degradation and the formation of inhibitory byproducts such as formic and levulinic acids.

Alkaline pretreatment processes are also being developed to deconstruct biomass by solubilizing lignin, as is done in pulping. Alkali recovery requires external energy input, and the high process temperature decreases sugar yield, as some hemicellulose is removed. Alkali recovery can be avoided by using ammonia for the pretreatment. This physicochemical process is known as ammonia fiber expansion (or explosion)—AFEX. Biomass is soaked in liquid ammonia at moderate temperature and high pressure. The ammonia penetrates the lignocellulosic biomass, causing it to swell while the ammonia reacts to alter the lignin structure and depolymerize cellulose. Rapidly reducing the pressure releases the ammonia, disrupting the fibers and leaving a substrate with much higher porosity that enhances the efficiency of enzymatic hydrolysis to give high sugar recovery ($\sim$99%). Ammonia vaporizes at the lower pressure, simplifying recovery, but process effectiveness decreases for higher lignin-containing biomass.

Hydrothermal processes use neutral pH physicochemical pretreatment with water at various temperatures and pressures. Near-boiling water at atmospheric pressure hydrolyzes cellulose and hemicellulose. This low severity process avoids degradation products, but the low concentration requires a large volume of water to recover sugars. Steam explosion is a combination of thermal, mechanical, and chemical methods. Saturated steam penetrates the fibers and expands the cell wall structure in much the same way as AFEX. Rapid pressure reduction leaves the porous structure intact for enzymatic hydrolysis. Steam explosion is very cost-effective. However, the high temperature can form inhibitory byproducts, and incomplete lignin removal lowers the sugar yield (impregnating the biomass with an acid catalyst improves the yield). Higher temperature and pressure hydrothermal pretreatment processes in subcritical and supercritical water have also been investigated but will likely be costly and challenging to operate at commercial scale.

Another promising pretreatment technology integrates mechanical and chemical methods. One approach combines chemical deacetylation and mechanical disk refining.[11] This is a two-step process: low temperature dilute alkaline deacetylation followed by disk refining with modest energy consumption. The result is a concentrated sugar hydrolysate with low levels of chemical and toxic inhibitors.

Hydrolysis

After pretreatment, any remaining polymeric or oligomeric carbohydrates (cellulose and hemicellulose) must be broken down into a monomeric sugar stream by hydrolysis. Several of the acid and hydrothermal methods discussed in the previous section only partly hydrolyze carbohydrates, depending on the severity of the pretreatment. A high-quality hydrolysate is required for optimum performance of the downstream fermentation, and increasing pretreatment severity to maximize hydrolysis is limited and needs to be controlled to avoid formation of inhibitory byproducts. Therefore, enzymatic hydrolysis is used to maximize monomeric sugar yield at modest process severity with the specificity of a biological catalyst to minimize inhibitory byproduct formation.

Separate biological catalysts (enzymes) are used to hydrolyze the remaining carbohydrates crystalline cellulose, amorphous cellulose, and hemicellulose. A suite of enzymes called cellulases have been developed to maximize hydrolysis of crystalline and amorphous cellulose. Specific enzymes target different regions of the crystalline polymer to reduce it to glucose. Amorphous cellulose is easier to degrade. Hemicellulose is typically amorphous and easier to hydrolyze than cellulose. However, it is a mixture of C_5 and C_6 sugars with different linkages that require different enzymes, xylanases, to degrade it into monomeric sugars. Of course, the challenge is to make sure that the products from one enzyme do not inhibit the activity of other enzymes.

Much work has been done over the past 20–30 years in the development of enzymes for cellulose and hemicellulose. Several companies offer cocktails that are sophisticated mixtures of enzymes and activities to effectively hydrolyze pretreated biomass to C_5 and C_6 sugars at high concentration and low enzyme cost.

Fermentation

Unlike the production of intermediate liquor by starch hydrolysis, which is pure C_6 glucose, biomass pretreatment and enzymatic hydrolysis yields a mixture of C_5 and C_6 sugars that require specialized fermentation microorganisms to maximize yield, with the most developed product to date being ethanol. Another complexity that arises for lignocellulosic sugar fermentation compared to starch-based fermentation is that the hydrolysate resulting from biomass pretreatment and enzymatic hydrolysis contains impurities that reduce or inhibit the productivity of fermentation organisms.

Several native yeasts and bacteria ferment glucose from starch to ethanol at high yields, rates, and titers. Natural organisms that ferment the mixture of C_5 and C_6 sugars with those same high yields still require development. Considerable biotechnology work over several decades to engineer organisms, both yeast and bacteria, to ferment hydrolysate has produced several engineered strains that work at good yields, rates, and titers. Examples are *Saccharomyces cerevisiae*, a yeast, and *Zymomonas mobilis*, a bacterium. Fig. 3.5 shows a pilot scale reactor for growing fermentation organisms.

As is often the case in complicated conversion processes, there is a tradeoff between capital cost and yield or efficiency. Lignocellulosic process conditions can be optimized independently for separate pretreatment, hydrolysis, and fermentation steps to improve yield and efficiency. However, this must be balanced against the higher capital cost. Several strategies for finding this optimal balance have evolved. These include a two-step process for separate hydrolysis and fermentation, single reactors for simultaneous hydrolysis and fermentation, and one-step consolidated bioprocessing.[12–15]

Fig. 3.5 Seed vessels for growing fermentation organisms. *(Photo by Warren Gretz, NREL 00943.)*

Product Recovery

Recovering ethanol is the most energy intensive step. Conventional distillation is used to concentrate ethanol in a near-azeotropic mixture (hydrous ethanol, which is about 95% ethanol and 5% water). The remaining water can be removed by molecular sieve adsorption to produce fuel (anhydrous) ethanol. In most lignocellulosic ethanol conversion processes, the lignin is used as the fuel source to provide the steam to drive the distillation process. Therefore, an external fuel source is not required, improving the carbon footprint of the process by keeping fossil CO_2 emissions low.

Thermochemical Conversion

These processes use thermal energy to dehydrate, devolatilize, depolymerize, and oxidize, partially or completely, lignocellulosic materials to produce heat and power, biofuels, and bioproducts. The methods can be arranged in terms of process severity, which is a function of temperature, pressure, oxidizer concentration, and residence time, to define pyrolysis, gasification, and combustion, as shown in Fig. 3.6. Increased temperature correlates with increasing oxidation for autothermal processes. Pressure does not necessarily affect severity but can affect product composition.

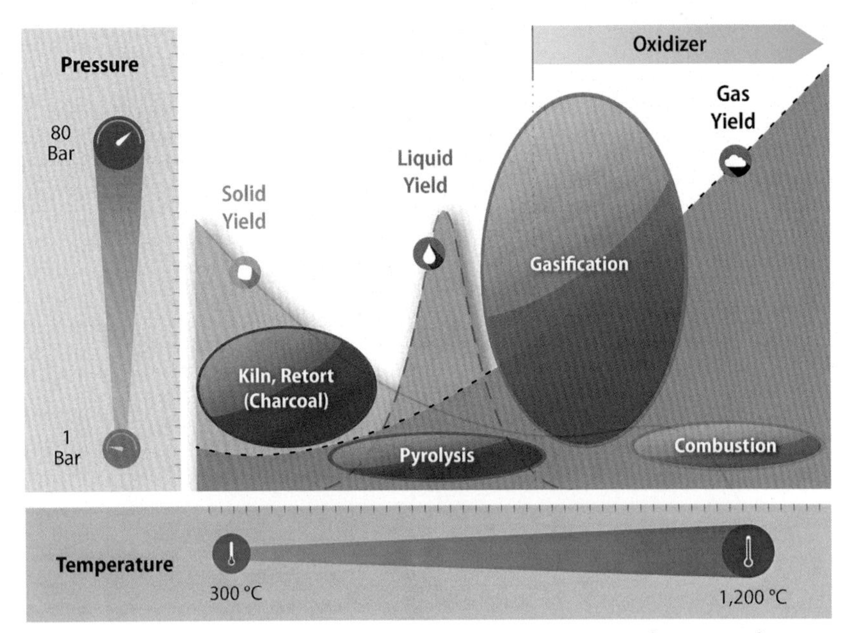

Fig. 3.6 Thermochemical biomass conversion processes as a function of process severity.

Pyrolysis

Pyrolysis is an option for producing liquid transportation fuels. It is the thermal depolymerization of biomass at modest temperatures in the absence of added oxygen. The slate of products depends on the temperature, pressure, and residence time of the liberated vapors. Flash (or fast) pyrolysis optimizes the liquid product, known as bio-crude or bio-oil. Bio-oils are multicomponent mixtures comprising different size molecules derived primarily from depolymerization and fragmentation reactions of the three key biomass building blocks cellulose, hemicellulose, and lignin.

Methods developed for crude oil are being adapted for bio-oil upgrading, with reasonable success to date. The primary challenge is that the elemental composition of bio-oil resembles that of biomass rather than that of petroleum fractions. The chemical and physical properties of crude bio-oil, such as high oxygen content, low pH, and poor thermal stability, are not directly translatable to petroleum processing. Therefore, recent research efforts have focused on (1) developing catalytic processes to modify the chemical and physical properties of bio-oil to better match petroleum processes for biofuel production, and (2) developing new processes and catalysts

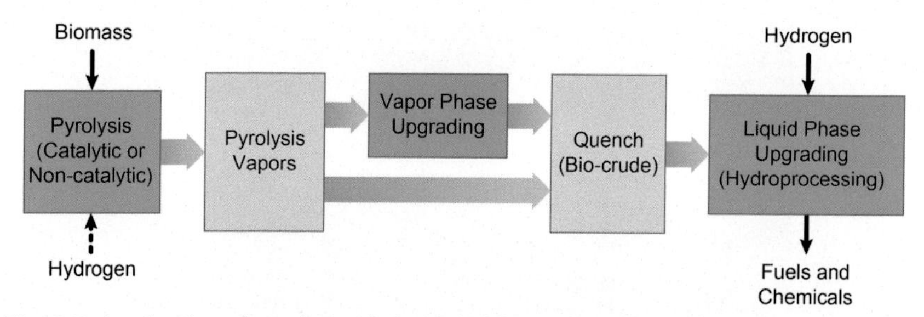

Fig. 3.7 Application of catalysis *(dark shaded boxes)* in biomass pyrolysis processes.

to upgrade bio-oils into biofuels. Fig. 3.7 is a summary of the options for various pyrolysis pathways with opportunities for applications of catalysts.

In biomass fast pyrolysis (FP), the material is rapidly heated (faster than 1000 °C/s) to 400–600 °C in an inert atmosphere (nitrogen) at ambient pressure with an inert heat transfer medium such as silica sand or silicon carbide. Vapor residence time in the reactor is typically <1 s to minimize vapor cracking and maximize liquid product yield.

If a catalyst is used instead of an inert heat transfer medium, the process is catalytic fast pyrolysis (CFP). It is sometimes called in situ catalytic fast pyrolysis to distinguish it from ex situ catalytic fast pyrolysis, or vapor phase upgrading, where the vapors are passed over a catalyst in a separate downstream reactor. If reactive gases such as hydrogen are added to the pyrolysis reactor in the presence of a catalyst, the process is reactive catalytic fast pyrolysis (RCFP). When biomass is pyrolyzed in the presence of a catalyst and pure hydrogen at elevated pressure (350–1000 psig), the process is hydropyrolysis (HYP).

The role of the catalyst is to promote deoxygenation of the vapors while minimizing carbon loss to char, light gases, and coke. Oxygen removal is done by dehydration (loss of H_2O), decarboxylation (loss of CO_2), and decarbonylation (loss of CO). Dehydration of the cellulose and hemicellulose fractions during pyrolysis, with or without a catalyst, produces water, referred to as water of pyrolysis, which is the most abundant single component of the liquid phase product. Biomass is inherently oxygen rich and hydrogen deficient, and the catalytic pyrolysis products become even more hydrogen deficient as dehydration occurs, which increases the tendency for aromatic formation and ultimately leads to char or coke production. Deoxygenation by CO and CO_2 removal (decarbonylation and decarboxylation) plus any carbon losses from coke formation on the catalyst lead to lower hydrocarbon liquid yield and lower energy recovery in the

Fig. 3.8 Dr. David C. Dayton in front of a 1-ton/day pilot plant in RTI International's Energy Technology Development Facility holding a jar of bio-crude produced from CFP of pine sawdust. *(Photo by Jimmy Crawford, RTI International.)*

bio-crude intermediate. Fig. 3.8 shows an example of bio-crude produced from pine sawdust in a CFP process.

The role of hydrogen during catalytic pyrolysis is to control coke formation on the catalyst surface and char production. Consequently, RCFP and HYP have emerged as potential solutions for maximizing carbon efficiency with improved hydrogen utilization.

Pyrolysis produces a liquid intermediate that can leverage years of technical development and capital expenditure in the petrochemical industry for upgrading into advanced biofuels. The degree of deoxygenation upstream has a large effect on the thermal stability of the intermediate. Lower quality bio-crudes may require a mild hydrotreating or stabilization step to make them more suitable for upgrading. Clearly the hydrogen consumption during the hydroprocessing step is directly related to the oxygen content of the intermediate. Upgrading bio-crude into transportation fuel requires removing oxygen and increasing the hydrogen-to-carbon ratio. Therefore, the central challenge in developing an integrated biomass pyrolysis bio-crude upgrading process is efficient utilization of hydrogen while maximizing carbon (energy) conversion efficiency and minimizing oxygen content in the product.

Gasification

Biomass gasification can be described as the partial oxidation of biomass in the presence of a gasifying agent, generally air, oxygen, or steam. Initial heating leads to evaporation of water. A further increase in temperature initiates pyrolysis, followed by the partial oxidation of pyrolysis vapors. The volatile matter that is released as the biomass fuel is heated partially oxidizes to sustain the endothermic gasification process. The char remaining after a biomass particle is devolatilized is also gasified. Fig. 3.9 shows the Joseph C. McNeil Generating Station in Burlington, Vermont that was the host site for the demonstration of an indirectly heated biomass gasifier fueled by wood chips. The product gas was sent back to the power plant and burned in the wood-fired boiler.

The product gas has low- to medium-energy content, depending on the gasifying agent, and is known as synthesis gas or syngas. It consists mainly of CO, H_2, CO_2, H_2O, N_2, CH_4, and other hydrocarbons. Gas composition and quality are dependent on a wide range of factors, including feedstock composition, type of gasification reactor, gasification agents, stoichiometry, temperature, pressure, and the presence or absence of catalysts.

Gas-phase impurities in syngas include NH_3, HCN, other nitrogen-containing gases, H_2S, other sulfur gases, and HCl. The concentration of these components strongly depends on the feedstock composition. Even low levels potentially threaten the successful application of downstream syngas utilization processes. Alkali metal, mostly potassium, is related to the alkali content in the biomass ash. Likewise, entrained ash particles affect the alkali metal content in the syngas. The concentration of alkali vapors or aerosols depends on the ash chemistry of the feedstock and the temperature of the gasification process.

Fig. 3.9 A view of the demonstration-scale indirectly heated biomass gasification unit at the Joseph C. McNeil Generating Station in Burlington, Vermont. *(Photo by Warren Gretz, NREL 04744.)*

The organic impurities range from low molecular weight hydrocarbons to high molecular weight polynuclear aromatic hydrocarbons. The higher molecular weight hydrocarbons are collectively known as tar. Tars are largely aromatic hydrocarbons with molecular weight greater than benzene, and tend to condense as syngas is cooled or compressed.

Syngas cleanup is a general term for removing the impurities. The individual contaminants can be managed separately, but collectively an integrated, multistep approach is required to reach acceptably low levels based on the end use. The most versatile approach is to cool the syngas stream and use a liquid scrubbing or absorption system. The thermal efficiency of the overall process is reduced because the syngas is cooled, the chemical energy in the tars is lost, and a waste stream is generated that needs to be processed or disposed of.

Alternatively, hot gas cleanup options, above the tar dewpoint, are being developed for tar conversion and impurity removal. In the absence of tars and particulates, there are commercially available sorbents and catalysts that mitigate catalyst poisons. Catalytic reforming mitigates tars; however, operating conditions must be carefully selected to avoid sulfur poisoning of reforming catalysts. The real challenge for managing all potential contaminants is determining which ones should be removed first and which can be dealt with collectively, and in what order based on process conditions. Ideally, an integrated process should have minimum capital and operating costs and maximum thermal and carbon efficiency.

Over the years, the gaseous mixture of CO and H_2 has had many names, depending on how it was formed: producer gas, town gas, blue water gas, synthesis gas, and syngas, to name a few. In the 19th century, coal gasification provided much of the syngas used for lighting and heating. Lignocellulosic syngas also can be converted to a variety of products by proven catalytic processes.

The beginning of the 20th century saw the dawn of fuels and chemicals synthesized from syngas. High pressure catalytic methods include methane production via the Sabatier process ($CO + 3H_2 \rightarrow CH_4 + H_2O$) discovered in 1902, ammonia production via Haber-Bosch synthesis ($N_2 + 3H_2 \rightarrow 2NH_3$) discovered in 1910, liquid hydrocarbon production (gasoline and diesel) via Fischer-Tropsch synthesis (FTS—$CO + 2H_2 \rightarrow —CH_2— + H_2O$) discovered in 1923, and methanol synthesis ($CO + 2H_2 \rightarrow CH_3OH$). Many of these routes were replaced with petroleum processes because of unattractive economics. In addition to hydrogen, methanol and ammonia continue to be made from syngas.

Catalytic synthesis dictates the optimum syngas composition, most importantly the H_2:CO ratio. That varies as a function of production technology and feedstock. Steam methane reforming yields a ratio of 3 to 1. Biomass gasification yields a ratio between 1.0 and 1.2, depending on the method, while coal gasification yields close to unity or lower. Consequently, syngas conditioning is used to adjust the properties of the syngas required for the catalytic synthesis process.

Except for steam methane reforming, catalytic syngas conversion processes are exothermic. Removing the large excess heat of reaction is paramount for controlling reaction temperature to maintain optimized process conditions.

Catalysts play a pivotal role in syngas conversion reactions. In fact, synthesis of fuels and chemicals from syngas does not occur without them.

Catalyst formulations have been optimized to contain the most active metals in combination with appropriate additives and supports to improve activity and selectivity in each process. In most of these processes, the number of moles of products is less than the number of moles of reactants, so they are more thermodynamically favorable at higher H_2 and CO partial pressures. These processes are also exothermic and generate large excess heat. Consequently, heat removal is critical for maintaining isothermal conditions to maximize product yields, minimize side or competing reactions, and maintain catalyst integrity and performance. As a result, reactor design and engineering has been an active area of research and development for temperature control and stability. Detailed process engineering and integration (with respect to heat integration and syngas recycle to improve conversion efficiency) are used to optimize commercial processes.

Biofuels can be produced from clean, biomass-derived syngas using FTS, methanol-to-gasoline technology, or a variant of FTS for mixed alcohol synthesis. FTS is a polymerization process initiated by dissociation of CO adsorbed on the catalyst surface followed by hydrogenation. Chemisorbed methyl species are formed by dissociation of absorbed CO molecules and stepwise addition of hydrogen atoms. These methyl species can further hydrogenate to form methane or act as initiators for chain growth. Chain growth occurs via sequential addition of CH_2 groups, while the growing alkyl chain remains chemisorbed to the metal surface at the terminal methylene group. Chain termination can occur at any time during the chain growth process to yield either an α-olefin or an n-paraffin once the product desorbs. The hydrocarbon products have a wide range of carbon numbers. Product distributions are influenced by temperature, syngas ratio, pressure, catalyst type, and catalyst composition. A common strategy is to optimize process conditions to maximize wax (high carbon number alkanes) production. Waxes are then upgraded to gasoline and diesel range hydrocarbons by hydrocracking.

Catalytic methanol synthesis is a classic high-temperature, high-pressure exothermic equilibrium limited reaction. Methanol can be used as a fuel or a blendstock, but it has more value as a commodity chemical. It can be converted to gasoline in a process that involves dehydration to produce dimethyl ether that can be converted to hydrocarbons over zeolite catalysts.

Mixed alcohols are a more attractive gasoline blendstock for octane enhancement, compared to ethanol and methanol. Mixed alcohols can be made from syngas by a variant of FTS that uses an alkali promoted catalyst to produce oxygenates instead of hydrocarbons. Unfortunately, commercial mixed alcohol synthesis is hampered by poor selectivity and low product yield.

Combustion

Combustion is the oldest method of biomass thermochemical conversion. It uses an excess of oxidant (air) to achieve the highest process temperatures by releasing the maximum amount of heat. Direct conversion in the form of fire was used for cooking, warmth, and light. Methods were developed to harness biomass energy to make electric power by indirectly using the heat of combustion to make steam that drives a generating turbine in a Rankine cycle. Continuous development has led to new technologies for biomass power systems with maximum thermal and electrical efficiency.

The simplest systems are wood stoves and furnaces for residential heating. Larger boiler systems are used for combined heat and power in commercial buildings and for distribution in local districts. Large systems are more fuel flexible but still have relatively low efficiency. Co-generation of heat and power improves thermal efficiency to as much as 85%. Boiler efficiency is affected by fuel moisture content, air-fuel ratio, excess air, combustion temperature, and biomass ash content.

Today, biomass power production is based on mature, direct-combustion boiler steam turbine technology. Improved control over the combustion process achieves better carbon conversion, higher efficiency, and lower emissions with more fuel flexibility. Most of the feedstock comes from the forest products industry (pulp, paper, timber, etc.) and is consumed internally for process heat. The power plants in commercial operation tend to be smaller (~20 MW_e) and less efficient (17–25%) than steam-turbine generators using coal or natural gas combined cycles.

Hydrothermal Liquefaction

A unique thermochemical method called hydrothermal liquefaction (HTL) uses hot, pressurized water to convert biomass into a viscous biocrude intermediate.[16] HTL is touted as an option for wet feedstocks such as food waste, wastewater treatment biosolids, paper mill sludge, animal manure, grain processing residues, macroalgae (seaweed, kelp), and microalgae without incurring the energy penalty that comes with dewatering and drying. Biomass slurries with 5–35% solids are pumped into the process. In some cases, such as microalgae, the process water demand is met by the feedstock, though dewatering is usually required for algae. Converting woody biomass or agricultural residues requires water to make the slurry; here, water recovery and reuse is essential.

The temperature can be adjusted to maximize solid, liquid, or gaseous products. Below ~250 °C the process is considered hydrothermal carbonization, with high char yield. At higher temperatures, pyrolysis reactions are initiated to maximize liquid biocrude yield. Above 450 °C, thermal cracking dominates in hydrothermal (or wet) gasification to maximize syngas yield. The pressure must increase as the temperature increases to keep water in the liquid state.

HTL is operated at ~250–450 °C and pressure at 100–350 bar, with residence time on the order of tens of minutes. These conditions are below the critical point of water, so it remains liquid. Even at subcritical conditions, the properties of water are changed significantly compared to ambient conditions. The density is about one-third and the dielectric constant 80% lower, so nonpolar organic compounds are now soluble. However, because the pyrolysis occurs in liquid water, ionic reactions that produce liquids are favored over free radical reactions that produce gases or solids.

The optimum temperature to maximize biocrude yield is strongly dependent on the feedstock.[17] Yields range between 17 wt% and almost 70 wt%. Regardless, the goal of HTL is to generate a biocrude intermediate with high energy density. The physical properties of this biocrude are different from those of biocrude produced by fast pyrolysis: It has roughly 10–15 wt% oxygen, with higher viscosity and lower density. The chemical compounds in the two biocrudes are similar, but the HTL type has more higher molecular weight components like naphthols and benzofurans, formed from the condensation reactions of smaller molecules.

Catalysts are used in HTL to increase biocrude yield while suppressing char formation. Na_2CO_3 and K_2CO_3 are the most common. They promote base-catalyzed condensation reactions. The alkali catalysts also dissociate into ions in the liquid water, enhancing the ionic mechanisms that favor liquid biocrude production.

The unique chemistry of biomass depolymerization in subcritical liquid water has attracted a lot of scientific attention and been the focus of many laboratory studies.[18] Extreme HTL conditions can be achieved quite routinely in small-scale laboratory batch reactors. Biomass, water, and catalyst are put into the reactor, and it is sealed, pressurized with an inert gas such as nitrogen, and heated to reaction temperature. The conditions are maintained for a fixed duration before it is cooled and depressurized so that the products can be collected and analyzed. Mass balances are accurate, because everything is contained in one vessel, and separating and analyzing different product streams provides detailed information for understanding the chemistry of the process.

Unfortunately, the temperature and pressure are not constant during these batch experiments, making it difficult to separate the effects of time and temperature and precluding determination of conversion or reaction rates. Batch processes are also not scalable for commodity chemicals and fuels. A continuous reactor system, in which temperature and pressure can be controlled independently, is required for future commercial HTL methods. The extreme conditions pose numerous technical challenges for scaleup and demonstration. There is not much commercial experience in pumping slurries with 5–35% solids into a high pressure, high temperature reactor. Reducing the pressure to collect the multiphase product and separating the viscous biocrude product from the aqueous and solid fractions is also a challenge. The heat transfer rate to heat the slurry and quench the products is another area that requires attention. HTL has the potential for making biocrude in high yield, but the multiphase reactants and products coupled with the severe process conditions add a great deal of risk for the demonstration and deployment of this technology on a commercial scale.

Hybrid Processes

Hybrid biomass conversion technologies have both biochemical and thermochemical steps. For example, heterogeneous catalytic processes for converting aqueous phase sugars into hydrocarbons have been developed. The overall process still requires a clean sugar stream generated from the pretreatment and enzymatic hydrolysis of lignocellulosic biomass but uses supported metal oxide catalysts to make aromatic hydrocarbons instead of yeast to ferment sugars to ethanol. Another hybrid under development uses gasification to deconstruct lignocellulosic biomass to produce syngas, then a biochemical conversion step to ferment the intermediate syngas to produce ethanol or other chemicals. Syngas quality is still a challenge, as impurities can adversely affect the fermentation organisms, but the process is less sensitive to syngas composition (H_2:CO ratio, CO_2 and N_2 concentration). Syngas compression is also not needed for the ambient pressure process, but optimizing mass transfer is critical for efficiency.

Close-up: Future Direction of Algae-Based Fuels and Products

The future bio-based fuels and products landscape will likely rely on a spectrum of different sources to meet the growing demand for replacements for petroleum-derived fuels and products. One of the most promising emerging sources is biomass derived from algae. These photosynthetic aquatic organisms can produce more biomass per acre than their terrestrial crop counterparts. It is envisioned that a higher value algal biomass-based industry will provide the additional revenue needed to reduce the net cost of producing algae-based biofuels.[19] A biorefinery approach that generates multiple high-value products will be essential to fully valorize algal biomass and enable economically viable co-production of bioenergy. To accelerate implementation, the primary objective needs to be maximizing productivity and product accumulation and minimizing energy, water, nutrient use, and land footprint of integrated algae-based operations, in particular for larger scale demonstrations and future research and development.[20, 21]

- There are tremendous opportunities associated with well-established bioenergy crops, and extending an existing agricultural framework to algae has great potential. When mapping out a trajectory toward the realization of a successful bioeconomy that includes algae, we need to be aware of remaining technical hurdles toward the economic viability of bioenergy.
- The single largest economic impact can be derived from improvements in biomass productivity. Progress here needs to be accompanied by energy, water, nutrient (fertilizer), greenhouse gas emissions, and land use metrics, with reliable techno-economic and life cycle analyses.
- Ecological, genetic, and biochemical development of strains is needed to improve productivity and robustness against perturbations such as temperature, seasonality, predation, and competition.
- Physical, chemical, biological, and post-harvest physiological variations of produced algae strains need to be researched and understood and integrated with biorefinery operations.
- On-site processing of algal biomass into its lipid, carbohydrate, and/or protein fractions needs to be developed at scales compatible with large-scale cultivation and commensurate with the respective markets.
- Recycling nitrogen, phosphorus, carbon, and other nutrients from residual materials after processing must be maximized to minimize the requirement for fresh fertilizer in cultivation.

For example, while it is possible to exploit the high photosynthetic efficiency of algae for bioenergy and biofuels production, inherent biological cellular constraints on strain production capacity are coupled to large differences in projected production scenarios for both micro- and macroalgae.[22] Still, two main areas are likely to become game-changing routes toward a

Close-up: Future Direction of Algae-Based Fuels and Products—cont'd

sustainable algae bio-economy: improving carbon capture and utilization to improve biomass productivity, and valuable bioproducts from process streams.

Large-scale algae deployment intersects biotechnology with agricultural innovations. Great strides are being made to develop agricultural approaches to maximize biomass yields and CO_2 capture. Deploying algae crop technologies benefits several metrics in terms of sustainability, thanks to the high photosynthetic efficiency and carbon assimilation potential, reducing the overall carbon footprint of biomass production. The 50% carbon in algal biomass, a conservative productivity of 20+ tons of biomass per acre per year, as demonstrated by nascent cultivation deployment, translates to an assimilation potential of over 40 tons of CO_2 per acre, by direct delivery of CO_2 to the ponds; this is at least double that of terrestrial crops. Current estimates based on national resource analysis place algae production potential at 104–235 million tons per year, with an integration of carbon capture and utilization.[23] This would make a significant contribution to national fuel output.

Ongoing research in carbon energetics engineering of both the algae and the agricultural cultivation approaches is increasing biomass productivity. For example, maximizing photosynthetic efficiency by improving the light harvesting systems or the carbon assimilation pathways has increased both cyanobacteria efficiency and terrestrial crop yields.[24–26] Advances in metabolic engineering tools applied to carbon energetics in algae have opened doors toward fundamental improvements in several areas, from algal biomass productivity[27] and product design[28, 29] to "hacking" photosynthesis in plants.[26] These recent discoveries indicate that the state-of-technology of algae processes and technical and economic feasibility might still be significantly short of the inherent biochemical potential, and future endeavors are likely to contribute to closing the gap between the theoretical and achievable potential for algae bioenergy.

Projected yields and productivity are highly dependent on the algae species (crop). Rapid crop rotation is used at the national testbed scale and at commercial farms.[30] Testbed experiments are designed to determine the feasibility and populate data to the underlying economic assessments for cultivation scaling, as well as the impact of innovation on farm deployment strategies. Opportunities are likely to accelerate in the near future with the inclusion of algae as a crop in the most recent farm bill. This may be a prelude to future financial and technical support from the US Department of Agriculture to help achieve the next frontier of large-scale deployment.

Continued

Close-up: Future Direction of Algae-Based Fuels and Products—cont'd

The emergence of algae on the bio-based products scene has prompted a redesign of biomass feedstock valorization. The decades of research that supported the feasibility testing of algae as biofuel feedstocks has given the field a platform to study the implementation of commercially viable opportunities.[19, 21, 31, 32] Thanks to sustained and strategic funding from the US Department of Energy's Bioenergy Technologies Office in the last 4+ years,[33, 34] progress has been made in the fundamental understanding of photosynthetic accumulation, outdoor deployment, and conversion of the biomass to fuels and products.[30, 32, 35, 36] Early commercial deployment of algae as a source of food and feed supplements is increasingly successful, and some products are now reaching supply chains in major brand supermarkets. These niche products, such as omega-3-fatty acid supplements,[37, 38] while perhaps unlikely to play a large role in the bioenergy arena alongside fuels because of limited market opportunities, do pave the way for the establishment of an industry that will benefit from transitioning to the commoditization of algae.

More than with any other currently cultivated crop, the composition of algae can be tuned to meet the demands of a changing market for producing fuels and a suite of products.[39–41] It is precisely this tunable biomass composition by species selection and environmental manipulation that informs a product suite. This unique contribution of algae, aligned with the appropriate conversion or fractionation technology, allows use of algal biomass components to drive down the cost of bioproducts compatible with a fuel-scale algae farm. A prime example is the recent push toward commercial production of polyurethane polymers from algae[42, 43]; given the high value and a recent drive to renewable polymers in a circular bioeconomy, this has great potential.

The interface between biotechnology and agriculture is set to position algae in a critical nexus to meet the increasing demands of biomass for fuels and products. The strategic direction for algae as feedstock can be achieved through the integration of biotechnology, process engineering, and analysis. Primary strategies for bioenergy production need to rely on a multi-product biorefinery approach to sustain economically feasible development. One of the products in the biorefinery would be fuels, produced alongside higher value products. The basic promise of algae-based bioenergy applications is still valid; there does not need to be competition with existing food and feed supply, since algae could be grown on non-arable land, with the added advantage of using wastewater as a cultivation medium to recover nutrients at each step of an integrated process that minimizes pressure on limited available resources. However, the other side of the coin is that there are significant barriers impeding commercialization and economical production for relatively low-value energy and fuel markets, in supporting the resource

Close-up: Future Direction of Algae-Based Fuels and Products—cont'd

demands for large scale deployment. The barriers range from incomplete knowledge of algae biology to challenges associated with the integration of technologies at demonstration scale. Even though many algae-based technologies have been demonstrated at laboratory scale, this most often has been focused on specific unit operations or aspects of the technology. The challenge remains to fully integrate algae-based processes and prove them out through extended multi-season operation.

Lieve M. Laurens
National Renewable Energy Laboratory, Golden, CO, United States

References

1. Ma F, Hanna MA. Biodiesel production: a review. *Bioresour Technol.* 1999;70(1):1–15.
2. Carlsson AS. Plant oils as feedstock alternatives to petroleum—a short survey of potential oil crop platforms. *Biochimie.* 2009;91(6):665–670.
3. Calero J, Luna D, Sancho ED, et al. An overview on glycerol-free processes for the production of renewable liquid biofuels, applicable in diesel engines. *Renew Sust Energ Rev.* 2015;42:1437–1452.
4. Guarnieri MT, Pienkos PT. Algal omics: unlocking bioproduct diversity in algae cell factories. *Photosynth Res.* 2015;123(3):255–263.
5. Ullah K, Ahmad M, Sofia, et al. Assessing the potential of algal biomass opportunities for bioenergy industry: a review. *Fuel.* 2015;143:414–423.
6. Brahic C, Venter C. One-minute interview renewable oilman. *New Sci.* 2009;203 (2718):25.
7. Galbe M, Zacchi G. Pretreatment of lignocellulosic materials for efficient bioethanol production. In: Olsson L, ed. *Biofuels.* 2007;108:41–65.
8. Agbor VB, Cicek N, Sparling R, Berlin A, Levin DB. Biomass pretreatment: fundamentals toward application. *Biotechnol Adv.* 2011;29(6):675–685.
9. Behera S, Arora R, Nandhagopal N, Kumar S. Importance of chemical pretreatment for bioconversion of lignocellulosic biomass. *Renew Sust Energ Rev.* 2014;36:91–106.
10. Rabemanolontsoa H, Saka S. Various pretreatments of lignocellulosics. *Bioresour Technol.* 2016;199:83–91.
11. Chen X, Shekiro J, Pschorn T, et al. A highly efficient dilute alkali deacetylation and mechanical (disc) refining process for the conversion of renewable biomass to lower cost sugars. *Biotechnol Biofuels.* 2014;7(1):98.
12. Lynd LR, Liang X, Biddy MJ, et al. Cellulosic ethanol: status and innovation. *Curr Opin Biotechnol.* 2017;45:202–211.
13. Rastogi M, Shrivastava S. Recent advances in second generation bioethanol production: an insight to pretreatment, saccharification and fermentation processes. *Renew Sust Energ Rev.* 2017;80:330–340.
14. Achinas S, Euverink GJW. Consolidated briefing of biochemical ethanol production from lignocellulosic biomass. *Electron J Biotechnol.* 2016;23:44–53.
15. Aditiya HB, Mahlia TMI, Chong WT, Nur H, Sebayang AH. Second generation bioethanol production: a critical review. *Renew Sust Energ Rev.* 2016;66:631–653.

16. Castello D, Pedersen TH, Rosendahl LA. Continuous hydrothermal liquefaction of biomass: a critical review. *Energies*. 2018;11(11):35.
17. Dimitriadis A, Bezergianni S. Hydrothermal liquefaction of various biomass and waste feedstocks for biocrude production: a state of the art review. *Renew Sust Energ Rev*. 2017;68:113–125.
18. Elliott DC, Biller P, Ross AB, Schmidt AJ, Jones SB. Hydrothermal liquefaction of biomass: developments from batch to continuous process. *Bioresour Technol*. 2015;178:147–156.
19. Laurens LML, Markham J, Templeton DW, et al. Development of algae biorefinery concepts for biofuels and bioproducts; a perspective on process-compatible products and their impact on cost-reduction. *Energy Environ Sci*. 2017;10(8):1716–1738.
20. Mussgnug JH, Klassen V, Schlüter A, Kruse O. Microalgae as substrates for fermentative biogas production in a combined biorefinery concept. *J Biotechnol*. 2010;150(1):51–56.
21. Ruiz J, Olivieri G, de Vree J, et al. Towards industrial products from microalgae. *Energy Environ Sci*. 2016;9(10):3036–3043.
22. Laurens LML, Chen-Glasser M, McMillan JD. A perspective on renewable bioenergy from photosynthetic algae as feedstock for biofuels and bioproducts. *Algal Res*. 2017;24:261–264.
23. Davis R, Coleman A, Wigmosta M, et al. *2017 Algae Harmonization Study: Evaluating the Potential for Future Algal Biofuel Costs, Sustainability, and Resource Assessment from Harmonized Modeling*. August 2018. NREL/TP-5100-70715.
24. Ungerer J, Wendt KE, Hendry JI, Maranas CD, Pakrasi HB. Comparative genomics reveals the molecular determinants of rapid growth of the cyanobacterium *Synechococcus elongatus* UTEX 2973. *Proc Natl Acad Sci*. 2018;115(50):E11761–E11770.
25. Xiong W, Morgan JA, Ungerer J, Wang B, Maness P-C, Yu J. The plasticity of cyanobacterial metabolism supports direct CO_2 conversion to ethylene. *Nat Plants*. 2015;1:15053.
26. South PF, Cavanagh AP, Liu HW, Ort DR. Synthetic glycolate metabolism pathways stimulate crop growth and productivity in the field. *Science*. 2019;363(6422):eaat9077.
27. Ajjawi I, Verruto J, Aqui M, et al. Lipid production in *Nannochloropsis gaditana* is doubled by decreasing expression of a single transcriptional regulator. *Nat Biotechnol*. 2017;35:647.
28. Radakovits R, Jinkerson RE, Darzins A, Posewitz MC. Genetic engineering of algae for enhanced biofuel production. *Eukaryot Cell*. 2010;9(4):486–501.
29. Work VH, Radakovits R, Jinkerson RE, et al. Increased lipid accumulation in the *Chlamydomonas reinhardtii sta7-10* starchless isoamylase mutant and increased carbohydrate synthesis in complemented strains. *Eukaryot Cell*. 2010;9(8):1251–1261.
30. McGowen J, Knoshaug EP, Laurens LML, et al. The algae testbed public-private partnership (ATP3) framework; establishment of a national network of testbed sites to support sustainable algae production. *Algal Res*. 2017;25:168–177.
31. Dong T, Van Wychen S, Nagle N, Pienkos PT, Laurens LML. Impact of biochemical composition on susceptibility of algal biomass to acid-catalyzed pretreatment for sugar and lipid recovery. *Algal Res*. 2016;18:69–77.
32. Dong T, Knoshaug EP, Davis R, et al. Combined algal processing: a novel integrated biorefinery process to produce algal biofuels and bioproducts. *Algal Res*. 2016;19:316–323.
33. Wigmosta MS, Coleman AM, Skaggs RJ, Huesemann MH, Lane LJ. National microalgae biofuel production potential and resource demand. *Water Resour Res*. 2011; 47(3):1–13. https://agupubs.onlinelibrary.wiley.com/doi/10.1029/2010WR009966.
34. Laurens LML, Slaby EF, Clapper GM, Howell S, Scott D. Algal biomass for biofuels and bioproducts: overview of boundary conditions and regulatory landscape to define future algal biorefineries. *Ind Biotechnol*. 2015;11(4):221–228.
35. Cano M, Holland SC, Artier J, et al. Glycogen synthesis and metabolite overflow contribute to energy balancing in cyanobacteria. *Cell Rep*. 2018;23(3):667–672.

36. Laurens LML, Nagle N, Davis R, et al. Acid-catalyzed algal biomass pretreatment for integrated lipid and carbohydrate-based biofuels production. *Green Chem*. 2015;17 (2):1145–1158.
37. Benemann J. Microalgae for biofuels and animal feeds. *Energies*. 2013;6(11):5869–5886.
38. Chauton MS, Reitan KI, Norsker NH, Tveterås R, Kleivdal HT. A techno-economic analysis of industrial production of marine microalgae as a source of EPA and DHA-rich raw material for aquafeed: research challenges and possibilities. *Aquaculture*. 2015;436:95–103.
39. Lang I, Hodac L, Friedl T, Feussner I. Fatty acid profiles and their distribution patterns in microalgae: a comprehensive analysis of more than 2000 strains from the SAG culture collection. *BMC Plant Biol*. 2011;11(1):124.
40. Laurens LML, Van Wychen S, McAllister JP, et al. Strain, biochemistry, and cultivation-dependent measurement variability of algal biomass composition. *Anal Biochem*. 2014;452:86–95.
41. Breuer G, Lamers PP, Martens DE, Draaisma RB, Wijffels RH. The impact of nitrogen starvation on the dynamics of triacylglycerol accumulation in nine microalgae strains. *Bioresour Technol*. 2012;124:217–226.
42. Kumar S, Hablot E, Garcia J, et al. Polyurethanes preparation using proteins obtained from microalgae. *J Mater Sci*. 2014;49:7824–7833.
43. Petrović ZS, Wan X, Bilić O, et al. Polyols and polyurethanes from crude algal oil. *J Am Oil Chem Soc*. 2013;90(7):1073–1078.

Additional Reading

Ail SS, Dasappa S. Biomass to liquid transportation fuel via Fischer Tropsch synthesis—technology review and current scenario. *Renew Sust Energ Rev*. 2016;58:267–286.

Bridgwater AV. Review of fast pyrolysis of biomass and product upgrading. *Biomass Bioenergy*. 2012;38:68–94.

Brown RC. *Thermochemical Processing of Biomass: Conversion into Fuels, Chemicals and Power*. Hoboken, NJ: John Wiley & Sons, Inc.; 2019.

Carpenter D, Westover TL, Czernik S, Jablonski W. Biomass feedstocks for renewable fuel production: a review of the impacts of feedstock and pretreatment on the yield and product distribution of fast pyrolysis bio-oils and vapors. *Green Chem*. 2014;16(2): 384–406.

Decker SR, Sheehan J, Dayton DC, et al. Biomass conversion. In: Kent JA, Bommaraju TV, Barnicki SD, eds. *Handbook of Industrial Chemistry and Biotechnology*. Cham: Springer International Publishing; 2017:285–419.

Foust TD, Ibsen KN, Dayton DC, Hess JR, Kenney KE. The biorefinery. In: Himmel ME, ed. *Biomass Recalcitrance*. Hoboken, NJ: Blackwell Publishing Ltd; 2009:7–37.

Karatzos S, van Dyk JS, McMillan JD, Saddler J. Drop-in biofuel production via conventional (lipid/fatty acid) and advanced (biomass) routes. Part I. *Biofuels Bioprod Biorefin*. 2017;11(2):344–362.

Mohan D, Pittman CU, Steele PH. Pyrolysis of wood/biomass for bio-oil: a critical review. *Energy Fuel*. 2006;20(3):848–889.

Rauch R, Hrbek J, Hofbauer H. Biomass gasification for synthesis gas production and applications of the syngas. *Wiley Interdiscip Rev Energy Environ*. 2014;3(4):343–362.

Sikarwar VS, Zhao M, Fennell PS, Shah N, Anthony EJ. Progress in biofuel production from gasification. *Prog Energy Combust Sci*. 2017;61:189–248.

CHAPTER FOUR

Analytical Methods in Biochemical Conversion

Introduction

The main goal in biochemical conversion of biomass is to design a process that uses biological catalysts, enzymes, and organisms to convert the polysaccharide portion—cellulose and hemicellulose—into high value fuels and chemicals. Since biological catalysts are used, C_5 and C_6 sugars are typically the desired intermediate. The lignin fraction, a phenolic polymer, is used for additional value generation, which can range from simple combustion for process heat to elaborate conversion processes for production of chemicals and materials.

Biochemical conversion processes use a wide spectrum of lignocellulosic biomass for fuels and chemicals, including crop residues such as wheat straw and corn stover, grasses such as switchgrass and fescue, woody feedstocks from both hardwood and softwood as well as forest residues, and dedicated bioenergy crops such as sorghum. The conversion efficiency and process economics are highly dependent on biomass composition, hence accurate feedstock analysis is critical. Particularly important is the ability to measure biomass carbohydrate content, because this is directly proportional to sugar yields and ultimately product yields. Minor components are also important; although they make up a small fraction of the biomass, they have significant impact on the yields and operability of a conversion process. They include ash, protein, organic acids, and nonstructural components. The ability to measure them accurately is necessary to provide mass closure.

There are many ways in which a biochemical process can be configured to make fuels and chemicals. The optimal configuration depends on the feedstock and the products desired. The flowchart in Fig. 4.1 is typical of a process for cellulosic ethanol and the analytical needs for each step.

Fig. 4.2 shows a configuration suggested by Chen et al.[1] for the production of a hydrocarbon biofuel and associated analysis needs. A major

Analytical Methods for Biomass Characterization and Conversion
https://doi.org/10.1016/B978-0-12-815605-6.00004-4

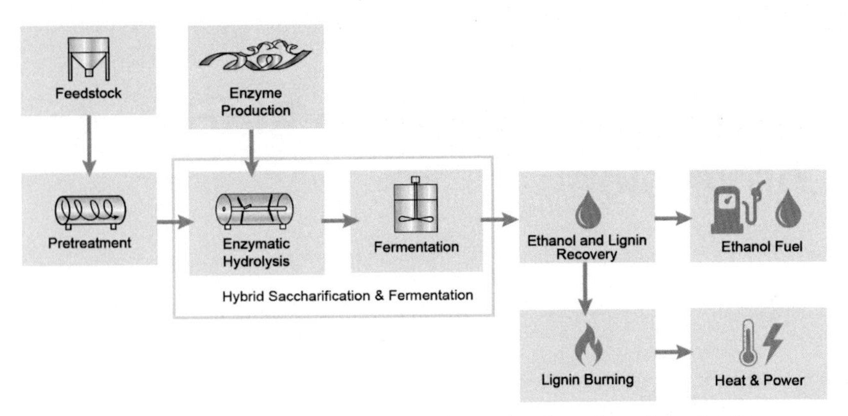

Fig. 4.1 Biochemical conversion process configuration for ethanol production.

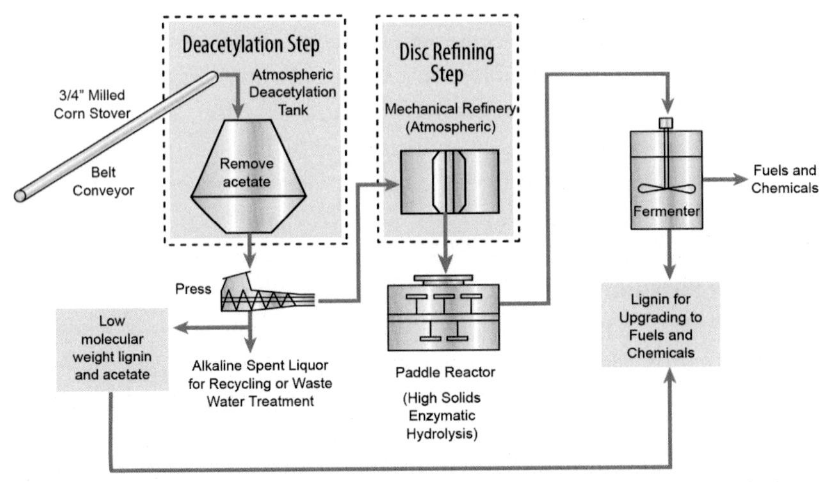

Fig. 4.2 Biochemical conversion process based on mechanical refining for fuels and chemicals production.

difference in the configurations shown in Figs. 4.1 and 4.2 is the saccharification for the liberation of the C_5 and C_6 sugars. Fig. 4.1 uses a chemical pretreatment process to liberate the C_5 sugars and condition the cellulose portion for enzymatic hydrolysis; Fig. 4.2 uses a NaOH deacetylation step followed by a mechanical disk refining step and then enzymatic hydrolysis. In the Fig. 4.2 process, total monomeric cellulose and xylose sugar yields were approximately 90% and 84% respectively, with a total sugars concentration of 150 g/L and low levels of the known fermentation inhibitors acetic acid, furfural, and 5-hydroxymethylfurfural.[2] This closely resembles the process shown in Fig. 4.1,with total monomeric and oligomeric cellulose and xylose sugar yields of 90% each and a total sugar concentration of 127 g/L in the hydrolysate, but with lower concentrations of inhibitors in the hydrolysate.[3]

Another significant difference between the two processes is the nature of the lignin and how it is depolymerized and collected. In Fig. 4.1, it is largely depolymerized and recondensed in the chemical pretreatment step. Although this is acceptable if the lignin is used as a fuel for heat and power production, the recondensed nature of the lignin makes it difficult for chemical production. In Fig. 4.2, the lignin component is largely depolymerized and liberated in the NaOH deacetylation step. Hence the degree of recondensement is greatly reduced, which makes the collected lignin stream much more amenable for chemical production. Since the two processes produce chemically different lignins, there are differences in analysis needs, which will be discussed in the next section.

Compositional Analysis Historical Development

Determining the composition of the input biomass as to structural carbohydrates, lignin, and trace nonstructural components is critically important to biochemical conversion. Of particular interest for biochemical production is the structural carbohydrate content, since this will determine the total amount of sugars available. This information is very useful in predicting yield potential and the economic value of the feedstock. The other critically important parameter is the trace compound content and the speciation of these compounds. Some of them can be detrimental to the conversion process. For example, ash reduces the efficiency and operability of several unit operations; hence knowing the ash content is important to determine the processability of a given feedstock. The total lignin content is useful as well; it determines the heating value of the lignin if it is used for heat and power generation as in Fig. 4.1, or chemical production as in Fig. 4.2.

Sulfuric acid hydrolysis has been used to determine lignin content and structural carbohydrates for more than 100 years and is the standard for compositional analysis. This method is very labor intensive, and, given the sensitivity to detail and technique, results can vary according to the person performing the analysis. Therefore great attention to detail is necessary when performing these analysis techniques. The method originated for wood lignin isolation, and Klason in 1906 is most often credited as the first to document it for standardization—hence the term "Klason lignin."[4] Subsequent researchers proposed improvements, all of which revolve around a two-stage design. The first stage is the use of a strong acid (65–72 wt% sulfuric acid) to reduce the structural rigidity of the wood into an amorphous material. The second stage is the use of a mild acid (0.5 wt% sulfuric acid) to remove the solids, then a second mild hydrolysis at elevated temperature (100–125 °C) to isolate the lignin. Because early lignin isolation and analysis was done by the pulp and paper industry, the methodological details were developed for woody feedstocks.

Early structural carbohydrate measurements were inconsistent. In 1933, Ritter et al.[5] found differences in hydrolysis time, acid loading, and temperature. They suggested an optimal set of conditions: 6 h at 16 °C or 2 h at 35 °C with a 4% sulfuric acid loading for the secondary hydrolysis stage to produce maximum yields for spruce. Slight subsequent adjustments provided a basis for standardization and consistency of results.

In the early 1930s, researchers[6, 7] applied the method to herbaceous feedstocks and also found inaccuracies, most likely because certain carbohydrates, proteins, or nitrogenous substances coprecipitated with lignin. Dunning and Lathrop[8] developed a modified hydrolysis procedure that, with later improvements, increased consistency.

The first application of these techniques to biofuel feedstocks was in the mid-1980s. Grohmann et al.[9] described a hydrolysis technique for wheat straw and reported difficulty in determining accurate glucose estimates. Subsequently researchers at the National Renewable Energy Laboratory (NREL) showed good component closure on a spectrum of herbaceous and woody feedstocks.[10] Of particular interest for biofuel conversion is the individual sugar concentration, so they introduced anion chromatography and pulsed amperometric detection (PAD) as well as high pressure liquid chromatography (HPLC). The HPLC methods require an additional ethanol extraction step prior to the sulfuric acid hydrolysis to extract carbohydrate fractions.

With the promise of a developing biofuel and bioproduct industry based on biochemical conversion, several improvements were made in the early 1990s to standardize compositional analysis so that results were repeatable and consistent. Milne et al.[11] published a comprehensive book on techniques, and it is a useful reference on the techniques of the time, but the authors did not perform comparison analyses across techniques. In 1992, the International Energy Agency (IEA) sponsored a round robin to test methods on four feedstocks being proposed for biofuel and biochemical production. More than 20 laboratories worldwide participated. HPLC and gas chromatography (GC) techniques were used to quantify carbohydrates. The results showed good consistency across the participating organizations and resulted in four National Institute of Standards and Technology (NIST) biomass feedstock Reference Materials (RMs): for sugar cane bagasse (RM 8491), eastern cottonwood (RM 8492), Monterey pine (RM 8493), and wheat straw (RM 8494).[12]

In the 1990s, there were several efforts toward intermediate stream and in-process stream analysis. Ehrman and Himmel[13] described differences necessary for in-process compositional analysis, specifically the need to separate pretreated slurries into a solid and a liquor fraction for separate analysis.[14]

Current Techniques

Feedstocks are heterogeneous, and their compositions vary significantly. Woody and herbaceous feedstocks differ from one another and within type. Corn stovers show up to 10% variation in structural carbohydrate and lignin content as a function of variety, harvest year, and geographical region. This variation has a significant impact on process economics and processability.

Considerable progress has been made in both wet and dry analytical techniques that give accurate repeatable results with full component mass closure and are well documented. Several process flowcharts have been proposed. Fig. 4.3, simplified from Sluiter et al.,[15] is a particularly good one. It shows in detail the laboratory analytical procedures (LAPs) that NREL provides for performing all of the steps. It summarizes the process and gives guidance on how to achieve mass closure as well as accurate, repeatable results. These LAPs are available at http://www.nrel.gov/bioenergy/biomass-compositional-analysis.html. Following is an overview of the processes and the information provided by the individual analyses.

Sample Preparation—Step 1

Sample preparation is absolutely critical for achieving accurate, truly representative results. Depending on the type of feedstock and how it was collected, it may contain dirt or other contaminants that need to be removed by a straightforward washing step. After contaminant removal come moisture reduction and size reduction. The LAPs specify that the sample should be $<10\%$ moisture to prevent microbial growth and degradation, and the analysis procedures require sample moisture below 10%. Air drying

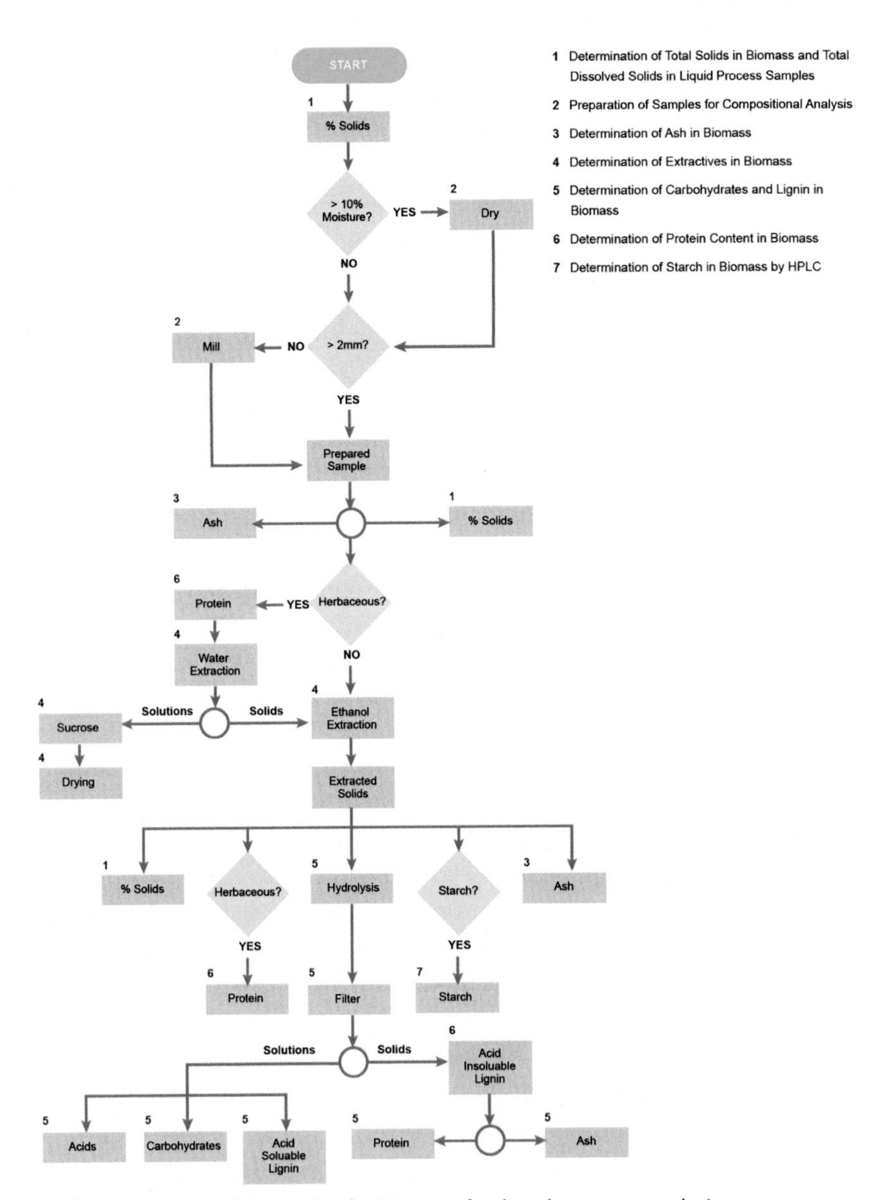

Fig. 4.3 Flowchart of the path of a biomass feedstock process analysis.

is the simplest if local conditions allow; if not, then a convection oven can be used. Drying temperature needs to be kept below 45 °C to prevent thermal degradation. More expensive techniques such as lyophilization or freeze-drying can be used for special cases such as high moisture feedstocks.

Size reduction is straightforward. A conventional knife mill is a good option for most feedstocks, woody or herbaceous. Some older protocols recommend sieving the milled sample with a 20, 40, or 80 mesh screen to remove smaller particles, because excessive degradation of smaller particles could occur during subsequent analysis, leading to distorted structural carbohydrate vs lignin content values. However, additional work has shown that removing the smaller particles leads to unrepresentative samples for herbaceous feedstocks, which tend to be heterogeneous, as the heterogeneity partitions by particle size. Hence most modern protocols do not recommend sieving and have modified downstream protocols to account for particle size heterogeneity. A seemingly trivial but important point is to be careful to homogenize the sample after size reduction and storage, because settling can cause inhomogeneity and an unrepresentative sample.

Total Solids—Step 2

Total solids content is typically reported on a dry weight basis to maintain consistency across feedstock types and moisture contents. For example, if a sample is 5% ash but contains 50% moisture on a dry weight basis, it would be 10% ash. Given that moisture content can be a factor for several parameters such as storage conditions and relative humidity in the analytical laboratory, accurately accounting for moisture content is important. The NREL LAP "Determination of Total Solids in Biomass and Total Dissolved Solids in Liquid Process Samples" outlines convective oven techniques and infrared moisture analyzer techniques.

Ash Content—Step 3

Inorganic matter is referred to as ash because that is how it is measured. Determining its content is necessary for mass closure and because high ash feedstocks can present challenges for conversion processes. The NREL LAP "Determination of Ash in Biomass" describes two highly accurate techniques using a muffle furnace.

Protein Content—Step 4

Protein is present in herbaceous feedstocks, primarily in leaves and stems. Measuring protein content is important if the food value of the feedstock needs to be known. Also, protein can interfere with downstream lignin analysis. Measurement is indirect: nitrogen content is measured, and then a conversion factor for the particular feedstock is used to determine the amino acid content. The NREL LAP "Determination of Protein Content in Biomass" gives details on how to conduct this analysis.

Extractives—Step 5

Extractives make up the nonstructural portion of biomass and are not chemically bound to the structural components carbohydrate and lignin. Their mass percentage varies by feedstock type and can be as high as 30% by weight. Like protein, extractives can be destroyed by downstream analysis and can interfere with some analysis processes, so they must be quantified and removed. NREL LAP "Determination of Extractives in Biomass" describes a water and ethanol extraction process and measurement of the extractives.

Starch Content—Step 6

In woody and mature dried herbaceous feedstocks, starch is minimal and does not need to be measured. However, if the feedstock contains grain or was harvested as an immature actively growing plant, starch content could be significant, and failure to account for it would artificially elevate the glucose measurement and distort the carbohydrate measurements. Several procedures exist, a particularly good one being the Megazyme Total Starch Assay.[16] Also, if the sample contains free sucrose or glucose, that should be removed by water washing to avoid artificial elevation of the carbohydrate measurements.

Structural Carbohydrate and Lignin Content—Step 7

The final and arguably the most important step is measuring the structural carbohydrate and lignin portion. This is a complicated process that requires great attention to detail. NREL LAP "Determination of Structural Carbohydrates and Lignin in Biomass" describes how to do it, and several organizations offer training on how to perform this LAP. It uses a two-stage sulfuric acid hydrolysis to measure total lignin, both acid soluble and insoluble, and a carbohydrate analysis by HPLC. The LAP describes preparation of all standards, process steps, HPLC calibration and setup, sugar recovery standards, and corrections necessary to achieve accurate results.

The technique for structural carbohydrate uses an anhydro correction factor based on monomeric sugar measurements. The process will not measure oligomeric sugars, so it is very important that all of the carbohydrates be in monomeric form. Sugar degradation is quite likely during the sulfuric acid hydrolysis; sugar recovery standards (SRSs) are used to account for it.

Lignin is measured in two separate fractions, acid insoluble and acid soluble. Acid insoluble (Klason lignin) is the high molecular weight portion. Determination of its exact nature requires additional analysis, such as by liquid chromatography-mass spectroscopy (LCMS), to assess the degree of polymerization and repolymerization and the nature of the intermonomeric linkages. Since both ash and protein are acid insoluble, the acid insoluble lignin values will need to be corrected; the LAP gives directions.

The acid soluble is the low molecular portion. The relative percentages of soluble and insoluble are feedstock dependent. The LAP describes an ultraviolet visible spectroscopy technique for determining the total acid soluble lignin portion and provides a list of extinction coefficients that will need to be employed, since they are feedstock specific. This analysis is very sensitive, and a number of interference factors such as high ash or high moisture may affect the accuracy of the results; the LAP describes a process to account for these interferences.

A very important point is that the process described in this LAP provides an accurate measurement of the total lignin content. However, these methods do not provide insight into the structure of the lignin. If a conversion process as depicted Fig. 4.2 is used where the lignin functions as a chemical conversion feedstock, the chemical and physical nature of lignin needs to be understood. Given that the chemical and physical nature of collected lignin is highly affected by pretreatment, in-process samples will need to be collected and analyzed. Several groups are developing techniques to determine the molecular weight, thus the degree of repolymerization, by determining the intermonomeric unit linkages as well as the physical aspects of the lignin stream. Most of these techniques use sophisticated LCMS protocols for making this determination. Unfortunately, at the time of writing, there is no LAP for how to perform this analysis.

Streamlined Techniques

Although the techniques described above are the most rigorous and will provide the most accurate results, sometimes it is only necessary to have compositional results to a first order degree of approximation. In this case, one needs a streamlined technique that doesn't incur the time and resource expenditures of the full process. NREL scientists have developed a technique based on near-infrared spectroscopy to predict the composition of a large variety of biomass types.[17] This method is correlated with data from the rigorous full compositional analysis techniques to develop models for a variety of biomass types, both woody and herbaceous.

Close-up: Advancing Bioenergy With Systems Biology

To power the planet in a more sustainable manner requires the development of new technologies and methods that use renewable feedstocks. One viable solution is a biological approach, taking advantage of the innate ability of living organisms to metabolize diverse feedstocks and yield an array of products. The substrates span the gamut of lignocellulosic sugars in their polymeric or monomeric forms, waste organic acids, and one-carbon (C1) compounds such as formate, carbon monoxide (CO), and methane. Even autotrophic microbes can convert CO_2 into CH_4 by using H_2 derived from electrolysis.

Every microbe is a collection of metabolic reactions operating in a network with a high degree of connectivity and coordination. The advent of genomics has enabled sequencing of genomes with minimal effort. A genome is the blueprint that encodes metabolic pathways. A systems biology approach entails the investigation of genomics coupled with transcriptomics and metabolomics to reveal quantitative expression of the various genes, intermediates. and products made by these pathways (Fig. 4.4). Collectively, this -omics-based approach is a powerful tool to advance our knowledge in constructing a native and subsequently a "rewired" metabolic network leading to targeted products. Yet one of the challenges is the underlying regulatory functions, including feedback mechanisms, controlling gene

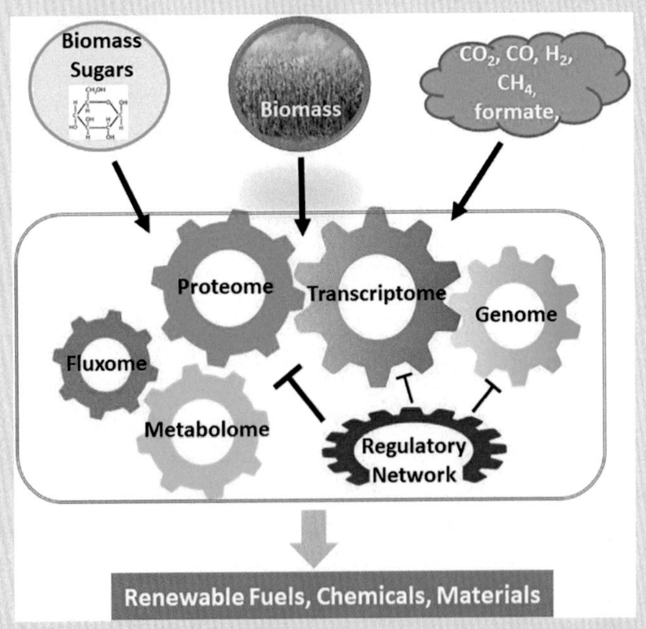

Fig. 4.4 A systems biology approach to convert renewable feedstocks to sustainable bioproducts.

Close-up: Advancing Bioenergy With Systems Biology—cont'd

expression or repression, which are often obscured until a pathway is perturbed. Gaining more knowledge of these dynamically regulated genetic circuits, from input signals to output responses, their targets, and the complex coordination afforded by the cells, is an emerging research area in order to fully realize the systems biology approach to redesigning microbial catalysts for targeted products.

A newcomer to the systems biology approach employs feeding microbes ^{13}C- or ^{15}N-tracer compounds and tracking their fate in a network of metabolic reactions. In ^{13}C-metabolic flux analysis (fluxomics), the cellulose-degrading bacterium *Clostridium thermocellum* was found to simultaneously fix CO_2 while metabolizing sugars.[18] This finding has important relevance in how certain microbes can recapture the CO_2 released from glycolysis, leading to higher carbon efficiency, which is an important premise for sustainable biofuels. This observation raises the questions how microbes manage and partition their carbon, energy, and electron flux to maintain redox homeostasis. Probing these design principles will guide cellular redesign toward more robust chassis strains. Maintaining energy and redox balance is especially crucial in CO_2-fixing acetogens (*C. ljungdahlii*, *C. autoethanogenum*) that must rely on H_2, CO, or formate as the C1 electron donor to build carbon-carbon bonds from CO_2 without photosynthesis.[19] Acetogenic microbes thrive at the thermodynamics limit of life, yet they are widespread in nature due to their metabolic flexibility. How energy is conserved in these acetogens is still an enigma, and systems biology could provide the breakthroughs to guide redesign of other microbes. Harnessing the power of microbial autotrophy and heterotrophy using a systems biology approach, including computational metabolic modeling, will therefore enable a new paradigm toward advancing bioenergy and biofuels production from sustainable resources.

Pin-Ching Maness

National Renewable Energy Laboratory, Golden, CO, United States

References

1. Chen X, Shekiro J, Pschorn T, et al. A highly efficient dilute alkali deacetylation and mechanical (disc) refining process for the conversion of renewable biomass to lower cost sugars. *Biotechnol Biofuels*. 2014;7(1):98.
2. Chen X, Wang W, Ciesielski P, et al. Improving sugar yields and reducing enzyme loadings in the deacetylation and mechanical refining (DMR) process through multistage disk and Szego refining and corresponding techno-economic analysis. *ACS Sustain Chem Eng*. 2015;4(1):324–333.

3. Humbird D, Davis R, Tao L, et al. *Process Design and Economics for Biochemical Conversion of Lignocellulosic Biomass to Ethanol: Dilute-Acid Pretreatment and Enzymatic Hydrolysis of Corn Stover.* Golden, CO: National Renewable Energy Lab. (NREL); 2011.
4. Klason P. Contributions to a more exact knowledge of the chemical composition of spruce wood, part I. *Pap Trade J.* 1922;74(18):45–51.
5. Ritter GJ, Mitchell R, Seborg R. Some factors that influence the conversion of cellulosic materials to Sugar[2]. *J Am Chem Soc.* 1933;55(7):2989–2991.
6. Peterson C, Walde A, Hixon R. Effect of temperature on sulfuric acid method for lignin. *Ind Eng Chem Anal Ed.* 1932;4(2):216–217.
7. Norman AG, Jenkins SH. The determination of lignin: errors introduced by the presence of certain carbohydrates. *Biochem J.* 1934;28(6):2147.
8. Dunning J, Lathrop EC. Saccharification of agricultural residues. *Ind Eng Chem.* 1945;37 (1):24–29.
9. Grohmann K, Himmel M, Rivard C, et al. Chemical-mechanical methods for the enhanced utilization of straw. In: *Paper Presented at: Proceedings, Sixth Symposium on Biotechnology for Fuels and Chemicals, Gatlinburg, Tennessee, USA, May 15-18, 1984*; 1984.
10. Vinzant TB, Ponfick L, Nagle NJ, Ehrman CI, Reynolds JB, Himmel ME. SSF comparison of selected woods from southern sawmills. *Appl Biochem Biotechnol.* 1994;45 (1):611–626.
11. Milne T, Brennan A, Glenn BH. *Sourcebook of Methods of Analysis for Biomass and Biomass Conversion Processes.* Springer Science & Business Media; 1990.
12. Lowenthal MS, Yen J, Bunk DM, Phinney KW. Certification of NIST standard reference material 2389a, amino acids in 0.1 mol/L HCl—quantification by ID LC-MS/MS. *Anal Bioanal Chem.* 2010;397(2):511–519.
13. Ehrman C, Himmel M. Simultaneous saccharification and fermentation of pretreated biomass: improving mass balance closure. *Biotechnol Tech.* 1994;8(2):99–104.
14. Wyman C. *Handbook on Bioethanol: Production and Utilization.* CRC press; 1996.
15. Sluiter JB, Ruiz RO, Scarlata CJ, Sluiter AD, Templeton DW. Compositional analysis of lignocellulosic feedstocks. 1. Review and description of methods. *J Agric Food Chem.* 2010;58(16):9043–9053.
16. Hall MB. Determination of starch, including maltooligosaccharides, in animal feeds: comparison of methods and a method recommended for AOAC collaborative study. *J AOAC Int.* 2008;92(1):42–49.
17. Sluiter A, Wolfrum E. Near infrared calibration models for pretreated corn stover slurry solids, isolated and in situ. *J Near Infrared Spectrosc.* 2013;21(4):249–257.
18. Xiong W, Lin PP, Magnusson L, et al. CO_2-fixing one-carbon metabolism in a cellulose-degrading bacterium Clostridium thermocellum. *Proc Natl Acad Sci.* 2016;113(46):13180–13185.
19. Schuchmann K, Müller V. Autotrophy at the thermodynamic limit of life: a model for energy conservation in acetogenic bacteria. *Nat Rev Microbiol.* 2014;12:809.

Additional Reading

Biomass Compositional Analysis Laboratory Procedures. www.nrel.gov/bioenergy/biomass-compositional-analysis.html.

Ehrman CI. Methods for the chemical analysis of biomass process streams. In: Wyman C, ed. *Handbook on Bioethanol: Production and Utilization.* CRC Press; 1996 [Chapter 18].

Milne TA, Brennan AH, Glen BH. *Sourcebook of methods of analysis for biomass and biomass conversion processes.* 1990. February SERI/SP-220-3548.

Sluiter A, Sluiter J, Wolfrum EJ. Methods for biomass compositional analysis. In: *Catalysis for the Conversion of Biomass and Its Derivatives.* 2013:vol. 2.

CHAPTER FIVE

Analytical Methods in Thermochemical Conversion

Introduction

A comprehensive understanding of the chemical composition of thermochemical biomass conversion intermediates and products is essential for process development and optimization. All inputs and outputs—solids, liquids, and gases—need to be collected and analyzed to close mass, energy, and element balances around a specific process to reduce performance uncertainty and maximize technical confidence. A versatile toolbox of methods and techniques can be applied to on-line monitoring of process streams while more traditional and standard analytical instruments can be used for post-process analysis of collected samples. In all cases, however, maintaining sample integrity during collection and storage and understanding the limitations of the methods and techniques are important.

Compositional analysis of the input biomass was discussed in Chapter 2. Many of these methods can also be used for analyzing solid intermediates and byproducts in gasification and pyrolysis. Solid residues from biomass gasification and pyrolysis are called char. On-line compositional analysis of char is typically not necessary, although char accumulation rates can be determined simply by weighing collected solids over time. Periodic sampling of solids and proximate, ultimate, and elemental ash analysis of collected samples can be used to determine carbon conversion efficiency of a thermochemical biomass conversion process. Monitoring trace contaminants, such as sulfur, nitrogen, chlorine, and alkali metals, that remain in the char is important for mitigating potential emissions from the internal use of char for process heat and for evaluating potential uses for excess char in energy or bioproducts applications.

Analytical Methods for Biomass Characterization and Conversion
https://doi.org/10.1016/B978-0-12-815605-6.00005-6

On-Line Process Analysis

On-line analyzers exist for monitoring permanent gases in thermochemical biomass conversion processes. In combustion systems, electrochemical oxygen sensors are used to continuously monitor excess oxygen to optimize combustion efficiency and minimize CO emissions. Continuous emission monitors (CEMs) for NO_x and SO_x are used for air permit compliance measurements. Non-dispersive infrared (NDIR) analyzers are also available for continuous monitoring of CO, CO_2, and CH_4. Real-time hydrogen monitoring can be done using analyzers to measure thermal conductivity.

Fourier transform infrared (FTIR) is another spectroscopic technique used for real-time monitoring of infrared active gas phase species. FTIR instruments can determine absolute concentrations of gas phase species from first principles (Beer's Law) without the need for calibration. Libraries containing FTIR data for many gas phase molecules are available for pattern matching to identify individual components in complex mixtures. Water is ubiquitous in many process gas streams and has spectral signatures across the near-, mid-, and far-infrared regions that can interfere with measurements of other gas phase components. Sophisticated signal processing algorithms are typically applied for selecting appropriate regions of the infrared range and deconvoluting overlapping signals from multiple species.

Rapid-cycle gas chromatographs, sometimes called refinery gas analyzers, are also available for measuring absolute concentrations of H_2, CO, CO_2, CH_4, and C_2-C_6 hydrocarbons every 3–5 min in biomass gasification and pyrolysis processes. Gas mixtures with known compositions are used to calibrate the retention times for specific components, while the area under the individual peaks correlates to absolute concentration. An example of the permanent gas analysis measured with an on-line gas chromatograph in a 1 ton/day catalytic biomass pyrolysis process is shown in Fig. 5.1. For perspective, RTI's 1 ton/day catalytic biomass pyrolysis unit is shown in Fig. 5.2.

On-line monitoring of condensable vapors is less common, but methods are available for analyzing volatile organic compounds. Residual gas analyzers (RGAs) have a capillary gas inlet to an electron impact ionizer that produces ions that are detected with a quadrupole mass analyzer. The capillary inlet is often heated to prevent condensation of water and other components and requires particulate-free gas samples so that the capillary does not plug. Gases with atomic mass between 2 and 300 amu can be analyzed

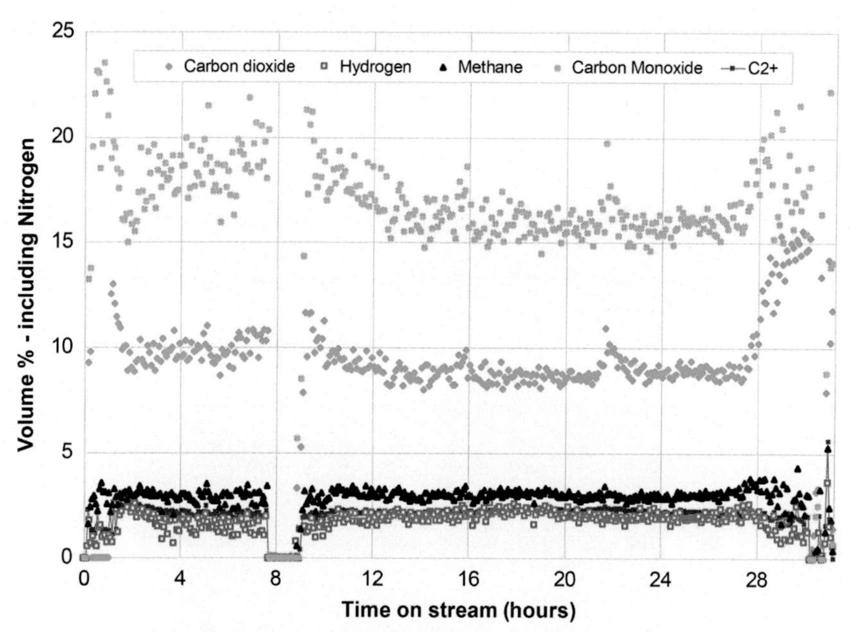

Fig. 5.1 Gas composition measured during catalytic pyrolysis of loblolly pine in a 1 ton/day circulating fluidized bed reactor.

with an RGA. The more complex the mixture, the more difficult the analysis, because individual components are usually not associated with a single mass signal. Electron impact ionization produces both a parent ion and fragment ions. Many gases have the same atomic mass; for example, N_2, CO, and C_2H_4 all have a mass of 28 amu. As a result, many components may not have a unique parent ion or fragment ions, causing overlapping signals. Fortunately, statistical data analysis and mass spectral library matches facilitate component identification from complex mass spectra.

Signal intensities in RGAs are dependent on several parameters, some associated with the instrument itself and others associated with the sample composition. Relative ionization cross sections of various molecules, ion optics voltage settings, and detector amplification are all instrument parameters that affect signal intensity. Consequently, calibrating RGA signals for absolute concentration is laborious and time-consuming and often not done. However, RGAs can be used for semi-quantitative analysis by comparing relative intensities of signals to added trace gases, typically argon. For example, normalizing mass spectral signals to the argon signal intensity can help to filter out instrument effects, making it easier to compare relative signal intensities of different components.

Fig. 5.2 The 1 ton/day catalytic biomass pyrolysis unit in RTI International's Energy Technology Development Facility. *(Photo by Jimmy Crawford, RTI International.)*

A unique on-line analysis method has been developed at the National Renewable Energy Laboratory (NREL) for sampling gasification tars, pyrolysis vapors, and other condensable vapor streams. It relies on molecular beam sampling as an alternative to a capillary inlet. Molecular beam methods have been used extensively in chemical physics to probe intermolecular potential energy surfaces by elastic, inelastic, and reactive molecular scattering. High resolution spectroscopy of weakly bound and hydrogen-bonded complexes formed in molecular beams with internal temperatures near absolute zero is an active area of research for studying how intermolecular forces can be used to understand condensed phases of matter. Molecular beam methods have also been applied to understanding the chemistry of flames. In particular, the detailed reaction chemistry and kinetics of mixtures of fuel and oxidizer has been investigated since the 1950s by using molecular beam sampling with mass spectrometric detection to measure concentration profiles of reactants, products, and reactive radical intermediates in low pressure (10–40 Torr) flames. In the early 1960s, Milne and Greene pioneered the use of molecular beam mass spectrometry (MBMS) for studying solid fuel combustion. This technique was brought to NREL (formerly known as the Solar Energy Research Institute, SERI) by Milne in 1977 and has since been used to study various aspects of biomass combustion, gasification, and pyrolysis.

MBMS is a three-stage differentially pumped vacuum system designed to sample process gas streams at high temperature (~400–500 °C) and ambient pressure. Process gases are sampled through a conical orifice on the order of 0.25–1.0 mm in diameter into the first stage of the vacuum system, creating a free jet expansion. The physics of this free jet expansion leads to a zone several orifice diameters downstream where intermolecular collisions are greatly reduced. In the absence of collisions, chemical reactions are quenched and condensation is inhibited. Therefore, the sampled gases and vapors are essentially frozen in their initial state.

The central core of this expansion is then directed through a conical skimmer with a 0.5–1.0 mm diameter and into the second stage of the vacuum system. By selecting the central core of the free jet expansion, a beam of molecules all moving in the same direction is created while the bulk of the sampled gases and vapors are deflected and pumped away. This molecular beam is then directed through another orifice into the third stage of the vacuum system that houses the quadrupole mass analyzer. The molecular beam is introduced into the ionizer at the entrance to the quadrupole mass analyzer, and ions are detected with an off-axis electron multiplier.

Analyzing mass spectral data collected with MBMS has the same challenges encountered with RGA sampling. Complex mass spectra require statistical data processing methods for identifying process trends, and only semi-quantitative information can be obtained when a tracer gas is used as an internal standard for normalizing signal intensities.

Another challenge with sampling condensable process vapors is maintaining sample integrity between the process and the analyzer. Often it is not possible to locate the analyzer close to the point in the process that is being monitored. This requires long sample transfer lines that must be maintained at process conditions (temperature and pressure) to avoid condensation, with minimal residence time to avoid further reaction of the sampled vapors.

An example of MBMS data collected during corn stover gasification in NREL's Thermochemical Process Development Unit (TCPDU)[1] is shown in Fig. 5.2. The TCPDU is a two-step gasification simulator with a 0.5 ton/day biomass feed rate. The first step is a bubbling fluidized bed reactor that operated up to 700 °C with a mixture of nitrogen and superheated steam input as the fluidizing gas. The partially oxidized vapors then flowed through a direct contact thermal cracker that could be maintained at temperatures between 500 °C and 900 °C. The gas was sampled downstream of the thermal cracker with a heated transfer line and directed to the MBMS inlet. The MBMS spectrum shows the traditional pattern for polynuclear aromatic hydrocarbons (PAHs), or tars. The high thermal cracker temperature produces tertiary tars; essentially PAHs that do not contain functional groups. The parent ion peaks for noted tertiary tars are labeled in Fig. 5.3. The peak at m/z = 40 is assigned to argon that is added as a tracer gas in a known volume. All ion signal intensities are normalized to the total argon ion counts.

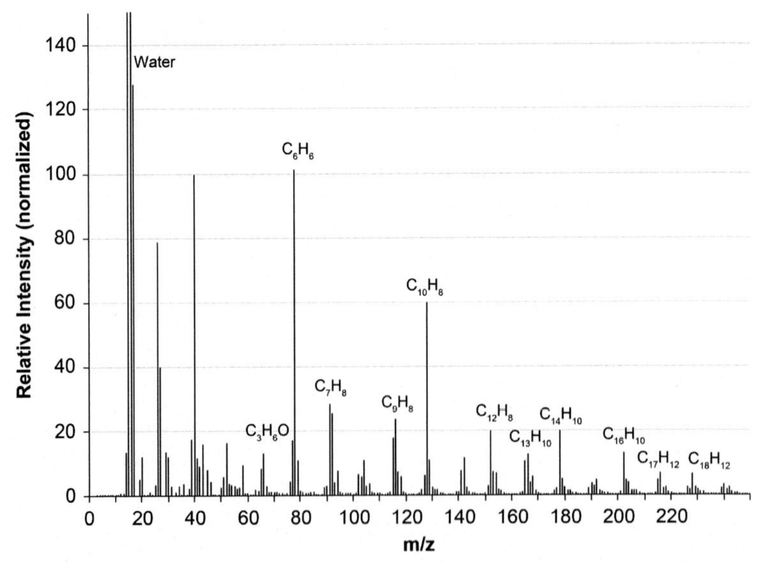

Fig. 5.3 Corn stover gasification in a 0.5 ton/day bubbling fluidized bed reactor at 650 °C with a steam:biomass ratio of 1. Syngas was thermally conditioned to 870 °C downstream of the gasifier reactor.

Using multiple on-line analyzers to sample different points within a process is usually cost-prohibitive. Therefore, if multiple points within a process are monitored, then multiple sample transfer lines are brought to a centralized manifold where switching valves direct the various samples to the inlet of the on-line analyzer. This becomes more challenging for condensable vapor sampling because of temperature limitations of selected hardware.

The techniques described have strengths and weaknesses depending on how they are implemented and used for on-line monitoring of a specific process. Unfortunately, a single method often does not meet the requirements for comprehensive chemical analysis, and the best approach requires multiple techniques with overlapping but complementary sample analyses. This requires a strategy to combine analysis methods to find a cost-effective solution that meets the needs of process monitoring and control. These comprehensive analyses are often validated with samples collected periodically and analyzed off-line during or after completion of the process.

Off-Line Sample Analysis

Getting representative samples from high temperature, high pressure thermochemical biomass conversion processes is often a challenge. For a more robust and comprehensive understanding of selected processes, on-line process monitoring is supplemented by extensive sampling periodically during an experiment or campaign or after an experiment. Once the sample is collected, a host of techniques and methods can be used to analyze it in much greater detail than might otherwise be done on-line.

Solids

Proximate analysis is used to determine moisture, volatiles, fixed carbon, and ash content of biomass and thermochemical intermediates. Measurements are done using a thermogravimetric analyzer (TGA) following ASTM D5142. This supports the measurement of material balance during an experiment or longer pilot-scale or production run. A typical proximate analysis is conducted by loading a small sample (~10 mg) into a TGA. The sample temperature is increased at a rate of 10 °C/min in helium to 105 °C. The weight loss is a measurement of the moisture content. The change in weight as the temperature is increased from 105 °C to 600 °C at a rate of 10 °C/min is a measurement of the volatiles content. At 600 °C, the gas is switched from helium to air. The change in weight is a measure of the fixed carbon in the sample that combusts. The residual mass left after oxidation is the amount of ash. A typical TGA profile for the proximate analysis of a solid sample is shown in Chapter 2 (Fig. 2.1).

Ultimate analysis is a measure of the elemental composition (CHON: carbon, hydrogen, oxygen, nitrogen) of biomass and thermochemical conversion intermediates and products. Samples are oxidized at high temperature (~1000 °C) in a furnace, and the combustion products (CO_2, H_2O, N_2, and SO_2) are separated on a column and analyzed by a thermoconductivity detector (TCD). In the case of oxygen determination, the samples are pyrolyzed in the furnace to form N_2, CO, and H_2. The CO is then separated from other gases, and the eluted gases are analyzed by the TCD. However, the oxygen content is calculated by difference.

Elemental analysis of process inputs and outputs supports the determination of element balances that can inform carbon efficiency measurements or energy efficiency calculations. Energy content can be measured directly by bomb calorimetry; however, correlations based on proximate and ultimate analyses have been developed with quite good accuracy. The following relationship, Eq. (5.1), can be used to calculate the higher heating value (HHV) of a sample based on the proximate and ultimate analyses[2]:

$$\text{HHV} = 3.55\text{C}^2 + 232 \times \text{C} - 2230 \times \text{H} + 51.2 \times \text{H} \times \text{C} + 131 \times \text{N} + 20{,}600 \tag{5.1}$$

where C, H, and N are mass percent of carbon, hydrogen, and nitrogen, respectively.

Proximate and ultimate analyses encompass the major components and elements in samples, but the trace components often affect processes in unforeseen ways. Trace elements like chlorine, phosphorous, alkali metals (Na and K), alkaline earth metals (Ca and Mg), and heavy metals (As, Hg, Ni, Pb, Cd, Se, Cu, and Zn) are of interest because they can be catalyst poisons or lead to adverse environmental emissions. Imaging solid particles with scanning electron microscopy or a variety of X-ray spectroscopic methods can identify surface metals. Inductively coupled plasma/mass spectrometry (ICP/MS) quantifies metals in solids.

Liquids and Condensable Vapors

Quenching product vapors from processes for liquid collection can be challenging. Heat exchangers, direct liquid contact spray quench, impinger trains, and filters are options. Measuring and quantifying condensable organics (tars) in biomass gasification processes is of interest. Tar yields in biomass-derived syngas depend on the gasification process and can range from 0.1 wt% (downdraft) to 20 wt% (updraft) or greater. The chemical composition of these tars is also a function of process conditions—temperature, steam:biomass ratio, equivalence ratio, and pressure.

Liquid sample collection and analysis is most relevant for pyrolysis processes. Liquid streams are defined as aqueous and organic fractions. Aqueous fractions are mostly water with up to a few percent dissolved organic compounds. Organic fractions are mixtures of organic compounds with a small amount of water. For example, bio-oil is an emulsion that can contain up to 20% moisture. Maximizing collection efficiency is the goal for understanding laboratory-scale experiments, optimizing pilot-scale integrated processes, or maximizing yields in larger demonstration and commercial units.

Tars

For the most part, tars are the condensable fraction of organic gasification products and are largely aromatic hydrocarbons, with molecular weight greater than benzene. The diversity in the operational definitions of tars comes from the variable product gas compositions required for end-use applications and how the tars are collected and analyzed. Sampling protocols have been developed[3–9] to help standardize collection methods, and new on-line analysis techniques are being evaluated[10,11] in addition to the MBMS sampling for on-line tar analysis in real time that was discussed above.

Other extractive sampling methods such as solid phase adsorption (SPA) are used for qualitative and quantitative analysis. A small slipstream of product gases is sampled through a cartridge containing the solid absorption material, commonly activated carbon and silica-based substances. Tar adsorbed on the cartridge is extracted by eluotropic solvents (i.e., solvents of increasing polarity). The column can be extracted with dichloromethane (DCM) to desorb nearly all aromatic hydrocarbons. Phenolic compounds not desorbed with the first solvent can be recovered with a DCM-isopropanol extraction. A third and final extraction can be done with isopropanol to ensure that all organic material has been recovered. The extracted liquid fractions are subsequently analyzed by gas chromatography/mass spectrometry (GC/MS).

A guideline for sampling biomass gasification tars was developed and tested from 1999 to 2002 as part of the International Energy Agency Bioenergy Gasification Task.[9] The system consists of a heated probe, a heated particle filter, a condenser, and a series of six impingers containing isopropanol. The impingers are sequentially cooled from about 20 °C (ambient) to −20 °C using ice and salt/ice baths. The gas is sampled for a specified period through the sampling line and filter. The volume, temperature, pressure, and gas flow rate through the sampling system are measured after the impingers. The total amount of tar collected is measured gravimetrically. The liquid samples are combined and the solvent is evaporated. The remaining organic liquid is weighed to determine the total amount of tar collected. The composition of each impinger sample is analyzed by GC/MS. The complex nature of the tars precludes quantification of all identified compounds, and the wide molecular weight distribution means that not all of the tars will elute through the GC column.

Bio-Oils, Bio-Crude, Upgraded Distillates, and Aqueous Fractions

Bio-oils are the product of biomass thermal decomposition. The chemical composition reflects the degradation products of cellulose (C_6 sugars), hemicellulose (C_5 sugars), and lignin (phenolics). The chemical complexity of bio-oils and bio-crudes is epitomized by a broad boiling point and molecular weight distribution of the components and thermal instability due to highly reactive components. Consequently, a single analytical method cannot comprehensively and quantitatively characterize these liquid products and intermediates.

Moisture content can be measured by Karl Fischer titration following the ASTM E203 standard test method for water using Hydranal-composite 5K reagent. Density measurements can be made with a portable density meter at 23 °C according to ASTM D4052. The kinematic viscosity of liquids can be measured with a 350 Cannon-Fenske upflow viscometer according to ASTM D-445-88.

Bio-oils and aqueous fractions also tend to have low pH. Total acid number (TAN) is a measurement, following ASTM D664, used for petroleum products to determine the naphthenic acid content of crudes to evaluate corrosivity of certain materials of construction. Bio-oil TAN is often reported even though it is typically two orders of magnitude higher than crude oils. The acidic components in bio-oils are carboxylic acids and phenolics. Modifications to ASTM D664 have been made to separate out the carboxylic acid contribution to the TAN.

An array of analytical methods, mostly chromatographic, are used to determine the chemical composition of bio-oils and bio-crudes. Gel permeation chromatography measures molecular weight distribution. Volatile components are analyzed by GC. Polar nonvolatile components are analyzed by liquid chromatography. Functional group analysis is done by FTIR and nuclear magnetic resonance (NMR).

GC/MS is the preferred method for compositional analysis of bio-oils, bio-crudes, aqueous phase samples, and upgraded products.[12, 13] The National Institute of Standards and Technology (NIST) mass spectral library is used for the initial identification of the most abundant compounds. Selected compounds are calibrated by using analytical standards prepared in accordance with ASTM D4307. For each compound, a linear calibration curve is established for selected concentration ranges to determine absolute concentrations of individual components.

Two-dimensional gas chromatography (GCxGC) has been used since 2000 to improve compound separation, peak capacity, and sensitivity.[14] It separates compounds that co-elute on a single column in one-dimensional GC. It uses two different columns, polar and non-polar, in series. Effluent from the first column is trapped for a fixed length of time before it is introduced onto the second column. Specialized software for data processing generates a two-dimensional chromatogram with polarity and boiling point on separate axes. Manipulating the injection and trapping times can separate and identify components in very complex mixtures. High-resolution mass spectrometry (time-of-flight and Fourier transform ion cyclotron resonance, for example) also facilitates compound identification.

Bio-oils and bio-crudes contain many oxygenated compounds that are quite reactive and cause thermal instability. This makes GC analysis difficult, as the samples are introduced onto columns at elevated temperatures. Derivatization is used to chemically alter polar compounds, like oxygenated organic hydrocarbons, to make them more thermally stable and less reactive so that they become more GC compatible. Silylation is used to replace active hydrogens with silyl groups. An example of a silylation agent is *N,O*-bis(trimethylsilyl)trifluoroacetamide (BSTFA). In BSTFA derivatization, the trimethylsilyl group replaces the active hydrogen in alcohols, acids, amines, amides, and thiols.[15]

^{13}C-NMR spectroscopy is used to evaluate the chemical composition of bio-oils, bio-crudes, and aqueous fractions. This avoids the issues of thermal stability and reactivity that limit GC. It analyses the whole sample dissolved in appropriate solvents. The carbon signals are categorized into eight main functional groups according to their chemical shifts (δ, ppm): aliphatic hydrocarbons (0–55); methoxyl carbon ($—OCH_3$) in phenolics (55–57); levoglucosan, anhydrosugars, alcohols, ethers (57–105); aromatic C—H

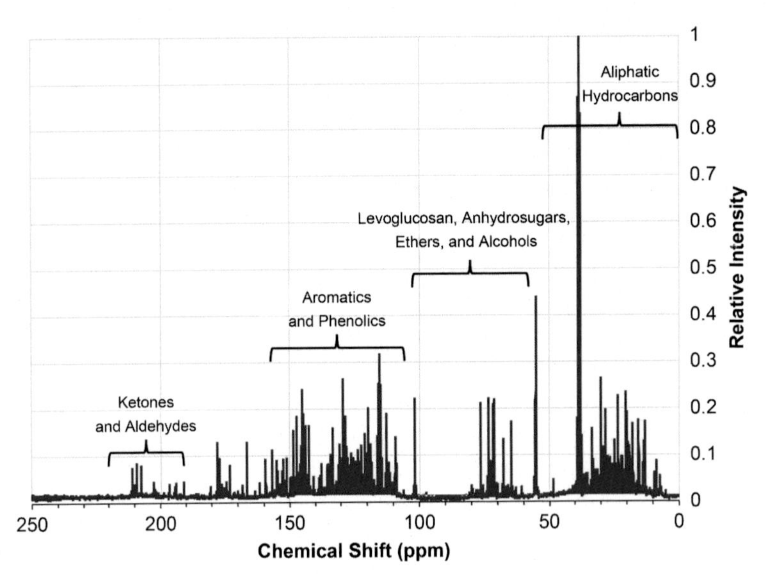

Fig. 5.4 ^{13}C-NMR of biocrude produced from the catalytic pyrolysis of loblolly pine at 527 °C.

bonds (105–125); aromatic C—C bonds (125–140); aromatic C—OH bonds (140–160); C=O in carboxylic acids and derivatives (160–180); and C=O in aldehydes and ketones (180–220). An example of a ^{13}C-NMR spectrum of biocrude produced from the catalytic pyrolysis of loblolly pine at 527 °C is shown in Fig. 5.4. The peaks associated with the representative functional groups are labeled in the figure. The signals from these chemical shifts can be integrated and the relative percent carbon content for each category can be determined. ^{31}P NMR for hydroxyl analysis is also useful in identifying different oxygen functionalities in bio-oils. The technique determines OH in aliphatic, phenolic, and carboxylic compounds.

Gases

On-line permanent gas sampling is preferred if analytical instruments are available. If not, process gases can be sampled and stored in vessels, tubes, pipes, or bags for post-process analysis by GC, FTIR, or other methods. Process gas can be flowed through sample vessels with valves on both ends to capture samples for off-line analysis. Another common method uses Tedlar™ bags that are made of chemically resistant fluoropolymer with low gas permeability. These bags are robust and low cost and often have a plastic or metal valve for easy sample collection.

Another off-line method is like the solid phase absorption techniques described above for tar. Activated carbon, silica gel, or molecular sieves are encapsulated in glass vials that contain a chemical reagent that reacts with a specific gas for a chemical family. Tubes are designed for specific concentration ranges of the gases being measured. A known amount of gas is drawn through the tube, and the color of the adsorbent changes along the length of the tube depending on the absolute concentration of the target gas in the sample. Tubes are available for ammonia, hydrogen cyanide, hydrogen sulfide, aliphatic (diesel and gasoline) and aromatic (BTEX) hydrocarbons, and chlorinated hydrocarbons. The accuracy and reproducibility of this method depends on how the gas is drawn through the tubes.

References

1. Carpenter DL, Bain RL, Davis RE, et al. Pilot-scale gasification of corn stover, switchgrass, wheat straw, and wood: 1. Parametric study and comparison with literature. *Ind Eng Chem Res*. 2010;49(4):1859–1871.
2. Friedl A, Padouvas E, Rotter H, Varmuza K. Prediction of heating values of biomass fuel from elemental composition. *Anal Chim Acta*. 2005;544(1):191–198.
3. Simell P, Stahlberg P, Kurkela E, Albrecht J, Deutsch S, Sjostrom K. Provisional protocol for the sampling and anlaysis of tar and particulates in the gas from large-scale biomass gasifiers. Version 1998. *Biomass Bioenergy*. 2000;18(1):19–38.
4. Knoef HAM, Koele HJ. Survey of tar measurement protocols. *Biomass Bioenergy*. 2000;18(1):55–59.
5. Maniatis K, Beenackers A. Tar protocols. IEA bioenergy gasification task. *Biomass Bioenergy*. 2000;18(1):1–4.
6. Xu M, Brown RC, Norton G. Effect of sample aging on the accuracy of the international energy agency's tar measurement protocol. *Energy Fuel*. 2006;20(1):262–264.
7. Xu M, Brown RC, Norton G, Smeenk J. Comparison of a solvent-free tar quantification method to the international energy agency's tar measurement protocol. *Energy Fuel*. 2005;19(6):2509–2513.
8. Horvat A, Kwapinska M, Xue G, Dooley S, Kwapinski W, Leahy JJ. Detailed measurement uncertainty analysis of solid-phase adsorption-total gas chromatography (GC)-detectable tar from biomass gasification. *Energy Fuel*. 2016;30(3):2187–2197.
9. Neubert M, Reil S, Wolff M, et al. Experimental comparison of solid phase adsorption (SPA), activated carbon test tubes and tar protocol (DIN CEN/TS 15439) for tar analysis of biomass derived syngas. *Biomass Bioenergy*. 2017;105:443–452.
10. Baumhakl C, Karellas S. Tar analysis from biomass gasification by means of online fluorescence spectroscopy. *Opt Lasers Eng*. 2011;49(7):885–891.
11. Defoort F, Thiery S, Ravel S. A promising new on-line method of tar quantification by mass spectrometry during steam gasification of biomass. *Biomass Bioenergy*. 2014;65:64–71.
12. Mullen CA, Boateng AA. Chemical composition of bio-oils produced by fast pyrolysis of two energy crops. *Energy Fuel*. 2008;22(3):2104–2109.
13. Cheng TT, Han YH, Zhang YF, Xu CM. Molecular composition of oxygenated compounds in fast pyrolysis bio-oil and its supercritical fluid extracts. *Fuel*. 2016;172:49–57.

14. Marsman JH, Wildschut J, Evers P, de Koning S, Heeres HJ. Identification and classification of components in flash pyrolysis oil and hydrodeoxygenated oils by two-dimensional gas chromatography and time-of-flight mass spectrometry. *J Chromatogr A*. 2008;1188(1):17–25.
15. Poole CF. Alkylsilyl derivatives for gas chromatography. *J Chromatogr A*. 2013;1296: 2–14.

Additional Reading

Black BA, Michener WE, Ramirez KJ, et al. Aqueous stream characterization from biomass fast pyrolysis and catalytic fast pyrolysis. *ACS Sustain Chem Eng*. 2016;4(12):6815–6827.

Ferrell JR, Olarte MV, Christensen ED, et al. Standardization of chemical analytical techniques for pyrolysis bio-oil: history, challenges, and current status of methods. *Biofuels Bioprod Biorefin*. 2016;10(5):496–507.

Michailof CM, Kalogiannis KG, Sfetsas T, Patiaka DT, Lappas AA. Advanced analytical techniques for bio-oil characterization. *Wiley Interdiscip Rev Energy Environ*. 2016;5(6): 614–639.

Milne TA, Abatzoglou N, Evans RJ. *Biomass gasifier "Tars": their nature, formation, and conversion*. Golden, CO: National Renewable Energy Laboratory; 1998. November 1998. NREL Report No. TP-570-25357.

Mohan D, Pittman CU, Steele PH. Pyrolysis of wood/biomass for bio-oil: a critical review. *Energy Fuel*. 2006;20(3):848–889.

Sampling and analysis of gases and vapours [Air monitoring methods, 2009]. In: *The MAK-Collection for Occupational Health and Safety*. Wiley-VCH Verlag GmbH & Co. KGaA; 2002.

Scoles G. *Atomic and Molecular Beam Methods*. Oxford University Press; 1988.

CHAPTER SIX

Analytical Methods in Hybrid Technologies

The primary biomass conversion technology options discussed in Chapter 3 involve chemical, biological, and thermal biomass deconstruction followed by biological or thermochemical upgrading of intermediates into biofuels and bioproducts. Biochemical and thermochemical process conditions are very different, and the physical properties and chemical composition of intermediate streams demand specific analytical methods. While several common measurements (temperature, pressure, pH, etc.) can be applied, unique analytical methods for process monitoring and stream characterization and compositional analysis are required for biochemical and thermochemical conversion.

Biochemical Processes

Biochemical processes involve biological catalysts at near ambient temperature and pressure in aqueous media. Understanding the physical and chemical properties of process streams requires specific analytic techniques that were described in Chapter 4. The relatively low concentration of analytes and products in water is a challenge for developing analytical methods. In situ methods are basically restricted to temperature and pH measurements. Ultrasensitive ex situ methods are available for liquid analysis, and optimization of biological catalysts (enzymes and fermentation organisms) is facilitated by DNA sequencing.

Biochemical catalysts can be engineered to transform reactants like lignocellulosic sugars or syngas into products with high conversion efficiency and minimal byproduct formation. However, biochemical processes tend to be very sensitive to impurities, and titers are often limited because the products can be toxic to the biological catalysts if the concentration is too high.

Analytical Methods for Biomass Characterization and Conversion
https://doi.org/10.1016/B978-0-12-815605-6.00006-8

Thermochemical Processes

Thermochemical processes involve high-temperature and in some cases high-pressure conditions using inorganic heat transfer media with or without catalytic activity to produce gases (gasification), condensable vapors (pyrolysis), or organic liquids (solvent and hydrothermal liquefaction). Analytical methods for monitoring thermochemical conversion and evaluating the physical and chemical properties of syngas and bio-oil intermediates were discussed in Chapter 5. The rather extreme process conditions make continuous monitoring of gaseous and vapor phase intermediates a challenge. However, condensed phase intermediates and products can be analyzed with selected ex situ methods. Methods and standards developed by the petroleum industry have been adapted for characterization of hydrocarbon-rich intermediates and products from thermochemical biomass conversion. Heterogeneous or homogeneous inorganic catalysts can be analyzed with proven high-resolution microscopy and spectroscopic techniques to understand structure-function relationships and catalyst deactivation.

Thermochemical processes are relatively insensitive to the input feedstock composition. However, there is limited control over the composition of the intermediate streams by applying inorganic catalysts and changing process conditions. Any impurities in the feedstock can be transformed into catalyst poisons that need to be managed downstream. Limitations of catalyst selectivity and difficulty maintaining isothermal process conditions for exothermic or endothermic upgrading reactions often produce a distribution of products rather than a single product molecule.

Hybrid Processes

Hybrid conversion technologies include a combination of biochemical and thermochemical steps within an integrated process configuration. The goal is to capitalize on the specificity of biochemical and the flexibility and robustness of thermochemical processes to maximize the carbon efficiency of developing advanced biofuels technologies. Within this context, the analytical methods described in Chapter 4 for biochemical conversion and Chapter 5 for thermochemical conversion can be applied to the appropriate unit operations in the hybrid integrated technology.

On-line process monitoring in hybrid systems is challenging because the type of unit operation requires expertise that is specific to either biochemical or thermochemical processing, depending on how the hybrid process is designed. There are different analytical paradigms that depend on the type of process involved, and not all engineers or operators may have the necessary breadth of experience or expertise with all methods. While there are no unique methods for hybrid technologies, a general familiarity with biochemical and thermochemical methods is required to identify which are appropriate to measure the most critical process conditions or stream compositions.

Laboratory analytical methods described in Chapters 4 and 5 can be applied to off-line sample analysis for both biochemical and thermochemical process intermediates and products. The choice of methods depends on the sample analyte concentration and substrate (solid, aqueous, organic, etc.), and standard methods can be applied when appropriate.

Conclusion

Hybrid technologies that integrate biochemical and thermochemical processes for advanced biofuels production do not require unique analytical methods for process monitoring, stream characterization, or product analysis. Those methods can be applied to the appropriate individual unit operations in hybrid systems. However, to maximize the carbon efficiency and yield of integrated hybrid technologies requires a general awareness of the analytical needs and challenges for both biochemical and thermochemical unit operations. Selecting the most important process variables to monitor and the most relevant intermediates and products to be analyzed supports the development of hybrid technologies that leverage the strengths of both biochemical and thermochemical.

Close-up: Advances in Hybrid Conversion Processes

With the need to convert a wide range of low-cost carbon feedstocks to displace petroleum, hybrid conversion processes provide the ability to integrate biological and chemo-catalytic unit operations to produce a wide array of renewable fuels and chemicals.[1–3] By using biological processes as the initial unit operation, microorganisms can leverage native and engineered metabolic pathways to convert a wide range of complex feedstocks (e.g., cellulosic sugars, industrial waste gas, lignin) into chemical intermediates with high-selectivity. By integrating subsequent chemo-catalytic unit

Continued

Close-up: Advances in Hybrid Conversion Processes—cont'd

operations, the throughput and diversity of accessible products can be expanded through carbon coupling reactions, oxygen removal, and incorporation of heteroatom functionality. Collectively, hybrid processes provide the ability to target drop-in petroleum products as well as unique molecular motifs for fuels and chemicals from low-cost carbon feedstocks.

Notable examples of hybrid processes are the production of renewable jet fuel and aromatic chemicals from fermentation-derived alcohols, both of which have been demonstrated at the precommercial scale. Alcohol-to-jet technology has been demonstrated in commercial airline flights, including fuel produced catalytically from fermentation-derived ethanol and isobutanol followed by chemo-catalytic dehydration and oligomerization.[4, 5] Alcohol-to-aromatics technology has been advanced for renewable para-xylene production from isobutanol.[6, 7] In regard to scalability, these processes use anaerobic biological methods because of the reduced aeration requirements that aid scale-up.[8] However, the high cost of conversion remains a significant barrier to commercial adoption of renewable drop-in fuels and chemicals. Moving forward, the ability to use low to negative cost waste feedstocks is increasingly important in order for biofuels and biochemicals to compete with existing petroleum products.

Recent efforts in hybrid processing have significantly expanded the diversity of waste feedstocks that can be converted to biological intermediates, as well as the diversity of drop-in fuels and chemicals produced catalytically. In the case of alcohol-to-jet, municipal waste has been used as the carbon source via gasification and syngas fermentation to ethanol, which applies a similar alcohol catalytic upgrading approach described above.[4, 5] This technology has been developed and scaled by LanzaTech, which recently completed a transatlantic flight on their waste-derived fuel with Virgin Atlantic.[9] Within the research domain, the number of waste feedstocks applied to hybrid pathways has seen significant growth. Lignin has long been viewed as a recalcitrant feedstock relegated to direct combustion for heat and power. Recent developments in hybrid processing have demonstrated biological funneling with a native strain of *Pseudomonas putida* to convert lignin-derived monomers into polyhydroxyalkanoates for subsequent catalytic processing into diesel-grade hydrocarbons.[10] To further expand the potential product slate from lignin, the strain was engineered to produce muconic acid as a key intermediate that can be catalytically upgraded to adipic acid and terephthalic acid.[11–15] Wet waste streams (e.g., sludge, animal manure, food waste) are a low-cost carbon source that benefits from existing waste collection and treatment infrastructure.[16] Mixed microbial consortia are able to convert complex waste organics into

Close-up: Advances in Hybrid Conversion Processes—cont'd

volatile fatty acids for further upgrading.[17] Recent developments in chemo-catalysis continue to expand the diversity of compounds that can be produced from short chain carboxylic acids, including alcohols, ketones, nitriles, and cyclic and branched hydrocarbons suitable for diesel and aviation fuel.[18–21]

Beyond the use of low-cost waste feedstocks, hybrid processing lends itself to the production of chemical compounds with unique structural or functional motifs that may provide performance differentiation from conventional petroleum products. From a fuel perspective, this includes targeting branched hydrocarbon structures that allow lower freezing points while retaining favorable autoignition properties for heavy duty vehicles.[22] In addition, oxygenated motifs (e.g., ethers, alcohols) can be incorporated to reduce the intrinsic sooting tendency of fuels for improved emissions, while retaining fuel performance properties.[22, 23] From a material perspective, molecules derived from hybrid processes have seen novel incorporation into polymer applications. For example, by incorporating β-methyl-δ-valerolactone, a monomer produced from the catalytic transformation of biologically derived mevalonate, the brittle nature of polylactide polymers can be improved to provide tunable mechanical properties.[24]

Moving forward, advancements in biology, separations, and catalysis will accelerate developments in this area. Utilization of CO_2 remains a grand challenge,[25] with new metabolic tools offering potential to improve carbon conversion efficiency, and emerging bioreactor designs aimed at addressing gas mass transfer limitations. The declining cost of renewable electricity also provides an entry point for electrochemical catalysis to merge with biological processes and afford selective chemical transformations that are difficult to access chemo-catalytically.[26–28] Cost-effective separation processes will also be key to isolating biological intermediates with sufficient purity for downstream catalytic upgrading.[29] Particularly for waste feedstocks, increased process intensity will be needed for industrial viability. In addition, the development of new catalyst materials that are robust and tolerant to high moisture environments and impurities can help reduce upstream separation burdens.[30] In short, significant opportunities remain for the future development of hybrid processes for the production of renewable fuels and chemicals.

Derek Vardon

National Renewable Energy Laboratory,
Golden, CO, United States

References

1. Goulas KA, Toste FD. Combining microbial production with chemical upgrading. *Curr Opin Biotechnol.* 2016;38:47–53.
2. Schwartz TJ, Shanks BH, Dumesic JA. Coupling chemical and biological catalysis: a flexible paradigm for producing biobased chemicals. *Curr Opin Biotechnol.* 2016;38:54–62.
3. Wheeldon I, Christopher P, Blanch H. Integration of heterogeneous and biochemical catalysis for production of fuels and chemicals from biomass. *Curr Opin Biotechnol.* 2017;45:127–135.
4. Geleynse S, Brandt K, Garcia-Perez M, Wolcott M, Zhang X. The alcohol-to-jet conversion pathway for drop-in biofuels: techno-economic evaluation. *ChemSusChem.* 2018;11(21):3692.
5. Wang W-C, Tao L. Bio-jet fuel conversion technologies. *Renew Sust Energy Rev.* 2016;53:801–822.
6. Collias DI, Harris AM, Nagpal V, Cottrell IW, Schultheis MW. Biobased terephthalic acid technologies: a literature review. *Ind Biotechnol.* 2014;10(2):91–105.
7. Maneffa A, Priecel P, Lopez-Sanchez JA. Biomass-derived renewable aromatics: selective routes and outlook for p-xylene commercialisation. *ChemSusChem.* 2016;9(19): 2736–2748.
8. Davis RE, Grundl NJ, Tao L, et al. *Process design and economics for the conversion of lignocellulosic biomass to hydrocarbon fuels and coproducts: 2018 biochemical design case update; biochemical deconstruction and conversion of biomass to fuels and products via integrated biorefinery pathways.* 11/19/2018. NREL/TP-5100-71949.
9. Topham G. First commercial flight partly fuelled by recycled waste lands in UK. *The Guardian.* 2018. US Edition.
10. Linger JG, Vardon DR, Guarnieri MT, et al. Lignin valorization through integrated biological funneling and chemical catalysis. *Proc Natl Acad Sci U S A.* 2014;111(33): 12013–12018.
11. Beckham GT, Johnson CW, Karp EM, Salvachúa D, Vardon DR. Opportunities and challenges in biological lignin valorization. *Curr Opin Biotechnol.* 2016;42:40–53.
12. Settle AE, Berstis L, Zhang S, et al. Iodine-catalyzed isomerization of dimethyl muconate. *ChemSusChem.* 2018;11(11):1768–1780.
13. Vardon DR, Franden MA, Johnson CW, et al. Adipic acid production from lignin. *Energy Environ Sci.* 2015;8(2):617–628.
14. Vardon DR, Rorrer NA, Salvachúa D, et al. cis,cis-Muconic acid: separation and catalysis to bio-adipic acid for nylon-6,6 polymerization. *Green Chem.* 2016;18(11): 3397–3413.
15. Lu R, Lu F, Chen J, et al. Production of diethyl terephthalate from biomass-derived muconic acid. *Angew Chem Int Ed.* 2016;55(1):249–253.
16. Milbrandt A, Seiple T, Heimiller D, Skaggs R, Coleman A. Wet waste-to-energy resources in the United States. *Resour Conserv Recycl.* 2018;137:32–47.
17. Lee WS, Chua ASM, Yeoh HK, Ngoh GC. A review of the production and applications of waste-derived volatile fatty acids. *Chem Eng J.* 2014;235:83–99.
18. Vardon DR, Settle AE, Vorotnikov V, et al. Ru-Sn/AC for the aqueous-phase reduction of succinic acid to 1,4-butanediol under continuous process conditions. *ACS Catal.* 2017;7(9):6207–6219.
19. Karp EM, Eaton TR, Sànchez I Nogué V, et al. Renewable acrylonitrile production. *Science.* 2017;358(6368):1307–1310.
20. Shylesh S, Gokhale AA, Sun K, et al. Integrated catalytic sequences for catalytic upgrading of bio-derived carboxylic acids to fuels, lubricants and chemical feedstocks. *Sustain Energy Fuels.* 2017;1(8):1805–1809.

21. Balakrishnan M, Sacia ER, Sreekumar S, et al. Novel pathways for fuels and lubricants from biomass optimized using life-cycle greenhouse gas assessment. *Proc Natl Acad Sci U S A*. 2015;112(25):7645–7649.
22. Fioroni G, Fouts L, Luecke J, Vardon D, et al. Screening of potential biomass-derived streams as fuel blendstocks for mixing controlled compression ignition combustion. *WCX SAE World Congress Experience*. 2019; https://doi.org/10.4271/2019-01-0570 April 9–11. Detroit, MI: SAE, Technical Paper 2019-01-0570. https://saemobilus.sae.org/content/2019-01-0570/.
23. Das DD, St. John PC, McEnally CS, Kim S, Pfefferle LD. Measuring and predicting sooting tendencies of oxygenates, alkanes, alkenes, cycloalkanes, and aromatics on a unified scale. *Combust Flame*. 2018;190:349–364.
24. Xiong M, Schneiderman DK, Bates FS, Hillmyer MA, Zhang K. Scalable production of mechanically tunable block polymers from sugar. *Proc Natl Acad Sci U S A*. 2014;111 (23):8357–8362.
25. Bushuyev OS, De Luna P, Dinh CT, et al. What should we make with CO_2 and how can we make it? *Joule*. 2018;2(5):825–832.
26. Matthiesen JE, Suástegui M, Wu Y, et al. Electrochemical conversion of biologically produced muconic acid: key considerations for scale-up and corresponding technoeconomic analysis. *ACS Sustain Chem Eng*. 2016;4(12):7098–7109.
27. Matthiesen JE, Carraher JM, Vasiliu M, Dixon DA, Tessonnier J-P. Electrochemical conversion of muconic acid to biobased diacid monomers. *ACS Sustain Chem Eng*. 2016;4(6):3575–3585.
28. Suastegui M, Matthiesen JE, Carraher JM, et al. Combining metabolic engineering and electrocatalysis: application to the production of polyamides from sugar. *Angew Chem Int Ed*. 2016;55(7):2368–2373.
29. Ramaswamy S, Huang H-J, Ramarao BV. *Separation and Purification Technologies in Biorefineries*. John Wiley & Sons, Ltd; 2013. https://onlinelibrary.wiley.com/doi/book/10.1002/9781118493441.
30. Schwartz TJ, O'Neill BJ, Shanks BH, Dumesic JA. Bridging the chemical and biological catalysis gap: challenges and outlooks for producing sustainable chemicals. *ACS Catal*. 2014;4(6):2060–2069.

CHAPTER SEVEN

Techno-Economics of Advanced Biofuels Processes

Introduction

Overcoming technical barriers is a key component of successfully developing and deploying advanced biofuels technologies. In the previous chapters, we presented the analytical methods used to identify and overcome technical challenges for a range of developing technologies. The underlying goal of these improvements is to reduce the cost of production of biofuels and bioproducts so that these alternative technologies become more economically competitive in existing markets compared to conventional petrochemical processes.

Techno-economic analysis (TEA) can be used to track the economic competitiveness of developing advanced biofuels technologies in parallel to the technical advancements. Alternatively, it can be used to highlight the technical barriers that have the largest impact on reducing costs to maximize the benefits of limited research and development resources.

Fig. 7.1 shows the methodology for conducting a TEA. It starts with developing a process design with process flow diagrams for individual unit operations. The fully integrated process includes unit operations for (1) biomass storage, handling, and preparation, (2) biomass conversion, (3) upgrading, and (4) product finishing. As much detail as possible should be included in the process flow diagram to make sure that no unit operation or major equipment is overlooked, so as much rigor as possible is incorporated into the capital cost estimates. At this point, a critical decision needs to be made defining the scale of the proposed commercial process. The scale defines the biomass resource needs and vessel and equipment sizes. Then models are developed to calculate material and energy balances. Cost estimation proceeds when the models are converged and all inputs and outputs can be calculated. Capital cost (CAPEX) is estimated based on the vessels and equipment in the detailed process flow diagrams. Operating cost (OPEX) is

Analytical Methods for Biomass Characterization and Conversion
https://doi.org/10.1016/B978-0-12-815605-6.00007-X

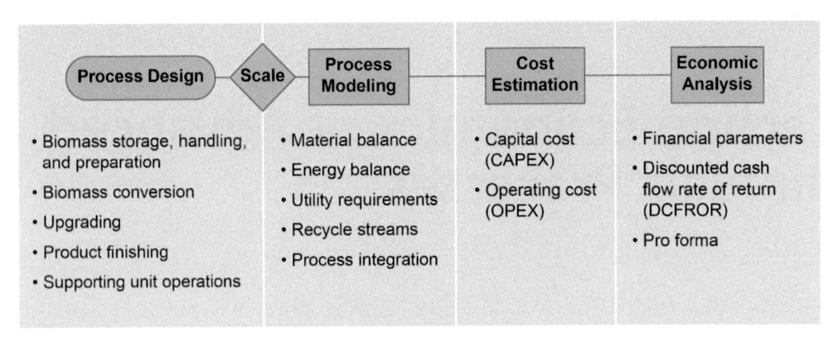

Fig. 7.1 Techno-economic analysis methodology.

based on the assumptions made for process operation (time-on-stream) and inputs and outputs calculated in the material and energy balances. Estimated CAPEX and OPEX and other financial parameters are inputs into a discounted cash flow rate of return (DCFROR) analysis to calculate a minimum fuel (or product) selling price (MFSP) or a pro forma analysis to evaluate economic scenarios where the integrated biofuels process is profitable.

It is exciting to consider the possibilities of developing new technologies based on scientific breakthroughs in the laboratory. Testing new catalysts or process conditions in carefully controlled laboratory reactors helps one understand detailed chemical reaction mechanisms and kinetics that support the development of new process concepts. Beyond basic proof of principle, the new process can be studied and optimized to maximize the yield of desired products, evaluate the impact of impurities or byproducts, and provide the foundation for additional research and development that supports technology scale-up and demonstration.

However, before significant resources and time are devoted to further process development and scale-up, it is prudent to evaluate technical feasibility and economic viability. Initiating a TEA very early in the timeline provides a baseline for tracking the economic impact associated with overcoming technical barriers and changes in market dynamics. An initial TEA often has a lot of uncertainty because bench-scale data is extrapolated to a commercial-scale, integrated process. Assumptions for yields, scale, and overall process performance from a limited amount of data under ideal laboratory conditions are also used for projecting the economics of a future n^{th} plant that assumes all technical barriers have been overcome in an optimized, integrated process.

The economic estimates for the initial process concept and the commercial plant define the endpoints with a clear target for a fully integrated, optimized commercial process. The TEA is continuously updated as technical barriers are overcome while development proceeds from the lab to the pilot scale to the demonstration scale. Over time, uncertainty in the TEA decreases as assumptions are replaced with experimental data, technical risk is better understood as unit operations are optimized, and economic factors are refined to reflect current market conditions as commercial viability approaches.

This chapter is about the role of TEA in advanced biofuels technology development. Important considerations for developing a robust TEA are highlighted, with more emphasis on the process of TEA than on specific economic results. Recognizing the impact of specific assumptions, technical and economic, is necessary when comparing TEAs for different processes. There are no commercial advanced biofuels processes to use as benchmarks, so comparisons to existing processes are necessary for defining specific assumptions based on industry practices. Ultimately, the relative economic results for different processes are more useful than comparing the absolute values for CAPEX and MFSP, for example, until advanced biofuels processes become a commercial reality and assumptions are validated and refined.

Process Design

Biomass conversion to fuels and chemicals has been an active area of research and development since the mid-1980s. Several technology options were described in Chapter 3. However, these processes are only part of an integrated biorefinery concept as the basis for a future commercial plant. Laboratory exploration and discovery are often focused on understanding or optimizing individual unit operations within the integrated biorefinery process. For example, biomass pretreatment is studied to optimize cellulosic sugar yields while minimizing byproducts and impurities. Experiments can then be designed to optimize the fermentation step. Similarly, syngas yields can be maximized in biomass gasification experiments, and upgrading experiments can be done to optimize catalytic syngas conversion to products.

Laboratory experiments are often done batch-wise. Different aspects of a continuous process can be studied in separate batch experiments. For example, switching from a reducing gas to an oxidizing gas can be used to investigate catalyst reduction and regeneration, and simulated distillation can be used to predict the performance of a distillation column for continuous product recovery.

Collectively, these experiments provide valuable information to define process conditions for individual unit operations, conversion efficiencies, stream compositions, and product yields that are the foundation for developing integrated biorefinery concepts. Detailed process flow diagrams for the biomass conversion and upgrading unit operations are developed with laboratory data as input to guide the design. Integration of unit operations is a key aspect of the system design that is not addressed in laboratory scale or pilot scale studies. Opportunities for heat integration and recycling of selected process streams are often critical for maximizing efficiency and minimizing operating costs.

Other individual steps not necessarily studied in the laboratory may need to be included. Sometimes these unit operations are based on commercially available technology, such as boilers for steam generation and natural gas reformers for hydrogen production. Options for individual unit operations can be based on commercially available technology that needs to be adapted for advanced biofuels production. The design may be well known, but assumptions about the performance in an integrated biorefinery must be developed and validated. Nevertheless, a robust design for an integrated biorefinery concept includes a comprehensive list of unit operations that are commercially-available, adapted from commercially available technology, or based on technology under development. The following areas are considered for fully integrated, stand-alone advanced biofuels technologies: biomass storage, handling, and preparation; biomass conversion; upgrading and product recovery; additional input requirements (air separation for oxygen and nitrogen, hydrogen production, enzyme production, etc.); steam and power generation; utilities (cooling water for heat exchange, caustic for neutralization and scrubbing, etc.); and wastewater treatment.

Scale

The overall throughput, or scale, of the integrated biorefinery concept needs to be defined before detailed process designs and equipment specifications are considered for each of the areas mentioned above. Capital costs do not typically increase linearly with equipment size. Therefore, one option for minimizing CAPEX per unit output is to maximize the plant size to take advantage of economies of scale. Obviously, there is a limit to plant size based on the capital investment required, but plant size is also limited by feedstock availability. The scale of advanced biofuels facilities is dictated by the availability of biomass resources that can be delivered to the plant at a given cost. At some point, biomass transportation and storage costs negate the economic gains for larger plant sizes.

Historically, 2000 dry metric tonnes per day (DMTD) has been used as the scale for advanced biofuels plants. This corresponds to corn stover collected from 10% of the land in a 50-mile radius around a plant located in the

Midwest region of the United States.[1] This plant size is large enough to take advantage of economies of scale yet small enough where biomass transportation distances are reasonable and biomass storage is manageable. Clearly, commercial opportunities and financial considerations dictate the scale of a given process that will be designed, built, and operated; however, 2000 DMTD is a mutually agreed scale in the research and development community for comparing different advanced biofuels technologies.

Biomass Storage, Handling, and Preparation

Biomass costs and logistics can vary as a function of feedstock and can be regionally specific. For example, the collection radius for dedicated energy crops like switchgrass or miscanthus can differ from corn stover based on higher yields and land availability for purpose grown energy crops. Seasonal harvesting of corn stover and switchgrass requires available storage for at least 9 months and typically 12 months. Woody biomass, on the other hand, can be "stored on the stump" and harvested from sustainably managed forests as needed continuously throughout the year.

A 2000 DMTD plant with a 96% capacity factor processes nearly 700,000 DMT of biomass per year. For agricultural residues like corn stover and grasses, the biomass is collected as bales. A standard square bale for storage is 1.2 m wide by 2.4 m long by 1.2 m tall and weighs approximately 0.5 tonnes. That means 1 years' worth of storage is 350,000 bales. If we assume that the bales can be stacked 10 high, and the area of each bale is 2.88 m^2, then the required storage area is 100,800 m^2, almost 25 acres. Clearly this will vary based on actual storage logistics and biomass bale density. If that amount of land area is not available at the plant site, biomass can be stored at satellite depots and delivered as needed. In this case, a minimum of 3 days' storage may be all that is needed, requiring a much smaller storage footprint but better access and logistics for biomass delivery. This also requires additional handling and transportation, adding to the cost of the delivered feedstock. The bottom line is that a lot of storage space is needed for a 2000 DMTD advanced biofuels plant, and feedstock logistics become important at commercial scale.

Ancillary equipment for feedstock handling and preparation also needs to be considered. Main equipment consists of conveyers; chippers, mills or grinders for size reduction; and dryers for thermochemical processes. Equipment selection will depend on the format (bales, chips, logs, etc.) of the delivered biomass and the biomass specifications for the envisioned process.

The scale of commercially available feedstock handling and preparation equipment also needs to be considered. Often a single piece of equipment is not available to handle 2000 DMTD of biomass, so redundant equipment is needed. Redundant equipment may also be built into biomass handling and preparation to meet reliability requirements. For example, a design may include three feed trains with the expectation that two will be operating continuously while the third is available as backup. If one of the two operating trains fails or requires unscheduled maintenance, the third train can be brought online without interrupting operation.

Biomass Conversion

Solids handling is one of the primary technical considerations for the scale and design of biomass conversion unit operations. The low bulk density and low energy density of biomass leads to larger scale conversion unit operations compared to technologies for gas or liquid processing. Also, reactor designs and equipment specifications are often unique to the selected process. Temperature, pressure, residence time, and other conditions influence reactor, vessel, and equipment size and materials of construction. Biochemical conversion requires vessels, tanks, pumps, etc. that are suitable for dilute aqueous phase processes that operate at low severity (temperature and pressure) and long residence time (hours to days), analogous to what is found in the corn ethanol and brewing industries. Thermochemical conversion processes operate at higher severity (temperature, pressure, and corrosivity) and shorter residence times (seconds to minutes). Reactors and equipment for thermochemical conversion are analogous to what is found in the petrochemical industry.

Practical limitations for commercial scale need to be considered when selecting equipment and sizing vessels. It is tempting to design a process that converts 2000 DMTD in a single tank or reactor to capitalize on economy of scale. However, holding tanks for biomass pretreatment or cellulosic sugar fermentation reactors may not be commercially available at that scale. Likewise, a single biomass gasifier or pyrolysis reactor can be designed for maximum throughput, but biomass feed systems at that scale may not be commercially viable. Processes should be designed with the appropriate number of duplicate or redundant trains and based on the scale of commercially available equipment and industry best practices for plant construction to achieve the design throughput.

Upgrading and Product Recovery

The liquid or gaseous intermediates produced during biomass conversion require additional processing. Upgrading and product recovery unit operations can usually be designed using proven, commercially available equipment with the scale matched to the throughput of the process area. Distillation columns for ethanol recovery and syngas conversion to fuels and chemicals are two examples.

Upgrading and product recovery may be based on commercially viable technology, but the application in an integrated biorefinery has not been demonstrated under commercially relevant conditions and timescales. For example, catalyst activity may be demonstrated for several hundreds of hours in laboratory experiments, but commercial catalyst replacement rates may require sustained catalyst performance for thousands of hours. Membrane separation for product recovery and concentration is another example where commercially available equipment can be applied without being fully validated under relevant conditions.

The technical risk of applying known technology for upgrading and product recovery is related to the potentially unique impurities introduced from biomass conversion. A critical process step that might be overlooked is one that mitigates the impact of impurities on upgrading. Impurities include solids or particulate matter, catalyst poisons, and byproducts. Laboratory or pilot-scale experiments are often not long enough to measure the impact of trace impurities, or the impurities can be present in such low concentrations that they are ignored or not detected. As technology advances toward commercial scale, these trace impurities must be dealt with to minimize impact on reliability, availability, and maintenance. Addressing these issues early in the technology development pathway can reduce future technical risk.

Utility Requirements—Gas, Electricity, and Water

Besides biomass as an input, what else is needed to run the process? Is it needed just for start-up or for continuous operation? When a given conversion process is scaled up from laboratory to a commercial size plant, it is easy to take certain inputs for granted. For biochemical conversion, enzymes and sulfuric acid needed for pretreatment may require a separate production unit or large storage tanks. For thermochemical conversion, oxygen-blown biomass gasification may require an air separation unit to provide oxygen, or a steam methane reformer may be needed to produce hydrogen for bio-crude upgrading. Other utility requirements include chilled water, fuel for heavy duty vehicles and generators, and other chemicals, catalysts, and materials.

Pumps, motors, compressors, and other equipment require electricity. Process streams like lignin from biochemical conversion or char from thermochemical conversion can be burned to raise steam for heat and power generation. This is an attractive option in terms of life-cycle greenhouse gas emissions reduction, because biomass is the original source of the carbon that is emitted. However, the capital costs for a power generation block can be a significant fraction of the total capital cost. Of course, electricity can also be purchased from the grid as an operating expense.

A similar scenario applies to water use and reuse. A wastewater treatment plant can be included within the boundary limits of the integrated biorefinery, or the process water can be discharged to a sanitary sewer if it meets the regulated requirements. Some form of wastewater treatment within an integrated biorefinery will likely be needed to maximize water recycle and minimize fresh water consumption.

Options are available when considering these additional process utility requirements. One is to include the production unit as part of the integrated biorefinery and add the capital costs as part of the total project investment. The additional cost of including these unit operations within the boundary limits is warranted if process intensity, emissions reduction, or environmental footprint is important. Another option is to assume that a vendor will provide the utility and the costs will be added to the total operating costs. This may be better suited for first-of-a-kind plants where it may be desirable to avoid the additional cost and risk of integrating multiple unit operations until the technology becomes more mature. Whether these utility requirements are added as additional CAPEX or OPEX, they will still be reflected in the fuel or chemical production cost.

Process Modeling

The robust design with detailed process flow diagrams for each area provides the basis for calculating material and energy balances for the integrated biorefinery. Experimentally determined conditions are defined in the design basis. Inputs are calculated from plant size, and outputs are calculated from the yields and materials balances measured in laboratory or pilot-scale studies.

Material balances are calculated for each area to validate or adjust equipment sizing. Compositions of the various process streams can be measured experimentally, but many streams are too complex to be used as input in modeling software. Also, physical property data for all stream components may not be available in software databases. Consequently, model or normalized process streams are created by matching the elemental composition of the experimentally measured process stream with a representative mixture of components that have been identified and are in the model database. For example, bio-crude is a mixture of several hundred compounds, and not all of them are in process modeling software databases. Therefore, representative hydrocarbons, aromatics, phenolics, carboxylic acids, aldehydes, ketones, and alcohols that are in databases can be combined to create a model bio-crude with an elemental composition that matches experimental data. Calculated element balances (carbon, hydrogen, oxygen, nitrogen, and sulfur) validates these estimated stream compositions and chemical reaction stoichiometries to make sure the chemistry is adequately represented in the model.

Thermodynamics and heat transfer requirements need to be satisfied so that temperatures defined in the design basis can be properly simulated by the model. The energy content of the product streams can be referenced to the input biomass energy content; however, heat can be generated by exothermic chemical processes, heat is required for endothermic chemical processes, and heat losses can occur. An energy balance across the system determines if additional fuel input is required to meet the heat energy demand, or if additional heat energy is available for export. Heat integration, especially in thermochemical conversions, is optimized to minimize the addition of external fuel and maximize energy efficiency. Heat integration can overcome heat loss, preheat certain streams, recover heat from heat exchangers used to cool those streams, provide heat for endothermic reactions, and use excess heat from exothermic reactions. Pinch analysis ensures that basic thermodynamics are not being violated by the heat integration.

The heat balance is one component of the energy balance. Electricity is also required to operate major equipment. A power generation block is included in the process design if there is enough high-quality excess heat available in the process to produce electricity. The power requirements of the major equipment will determine if external electricity is needed or if excess electricity can be exported for an additional revenue stream.

With the material and energy balances closed, the fully validated process simulation can be exercised with confidence to determine an overall energy efficiency and carbon conversion efficiency based on biomass input, water consumption, and any additional external fuel needed. The costs for the inputs and outputs from the model will be used to estimate operating costs in the economic analysis.

Cost Estimation

CAPEX

A comprehensive equipment list can be generated from the process flow diagrams for the individual process areas. The most accurate method for determining equipment costs is to get quotes from vendors, but this is not very common, especially for research and development projects. For specialized equipment such as custom designed reactors and vessels, the process conditions define the necessary materials of construction. The cost of custom equipment is determined from the material cost and the amount of material needed for the vessel adjusted by a scale factor for estimating the fabrication cost. For standard equipment items such as pumps, heat exchangers, and separation columns, databases are available with cost information that can be scaled to the size needed.

The total estimated cost of all identified equipment defines the total purchased equipment cost. An installation factor is then applied to the purchased equipment costs to estimate the total installed equipment cost. The installation factors take into consideration instrumentation and controls, process piping, electricals, and civil and structural costs. Installation factors are based on industry standards. The sum of all these factored costs provides an estimate for the total installed cost (TIC).

Additional factors are used to estimate the indirect costs that include engineering, construction, project management, legal and license fees, and contingencies. Applying these factors to the TIC yields the total project investment (TPI). All capital costs can also be indexed to the proposed project year to normalize costs to global macroeconomic factors such as economic growth and inflation.

OPEX

Operating costs are separated into variable and non-variable. Variable costs are incurred only when the process is operating and include those for feedstock, catalyst, other raw materials and chemicals, waste disposal, electricity, and makeup water. Any co-product credits or subsidies reduce variable operating costs. Fixed operating costs are incurred whether the process is operating or idle, and include labor, overhead, income tax, insurance, and capital depreciation.

Economic Analysis

The detailed process design and process model are used to develop robust capital and operating cost estimates. A simple fuel or product production cost can be calculated by dividing these costs by the modeled product yield. However, a more sophisticated economic analysis includes the total project investment and fixed and variable operating costs, plus financial and economic factors, to calculate the net present value of the total project as a function of the internal rate of return. The economic model can be a pro forma with a defined fuel or product selling price used to determine the rate of return (profit) or a discounted cash flow rate of return (DCFROR) with a fixed rate of return based on a calculated minimum fuel or product selling price to achieve that return on investment. In short, the economic analysis seeks scenarios where the technology can be profitable.

A pro forma or a DCFROR requires several economic and financial assumptions. For example, the amount of equity financing can be defined with the remaining debt financed with loans at a specific interest rate for a given term. A depreciation method is specified for a given plant life and a defined construction and start-up duration. Federal corporate income tax rates are calculated based on annual volumetric production and capital depreciation. The location of the plant will define the state income taxes, but these are typically not included for TEAs of advanced biofuels technologies reported in the literature, because locations are usually not defined for research studies.

A sensitivity analysis around the various capital and operating cost components can be performed to identify where future technology development resources and efforts should be focused to have the biggest impact on the process economics. Non-technical economic parameters such as debt financing rate, fuel selling price, and the size and availability of subsidies often have large impacts on the biofuels process economics but are not within the control of technology developers. Process improvements that reduce feedstock cost, improve yield, extend catalyst lifetime, reduce process severity, and maximize energy efficiency also have direct impact on the process economics and can be affected by technical advancements. Incremental progress toward intermediate technical targets can be used to track economic improvements on a regular basis to establish the current state of technology for a given advanced biofuels pathway. This validates the investments for research and development and defines the trajectory toward a commercial technology.

The economic potential of a given advanced biofuels technology once it is fully developed can be evaluated based on what is referred to as "n^{th} plant" economics. This case assumes that the technical targets have been achieved and the milestones met for process improvements. The nth plant economics assume that the technology has reached commercial maturity—the capacity factor (time on stream) is maximized and the reliability, availability, and maintenance (RAM) schedules are fully developed. Without specific examples of commercially operating plants, commercial maturity is based on assumptions for future technology development and deployment. Regardless, if assumptions are consistent, the nth plant economics provide a relative comparison of the economic potential of different technology options and an evaluation of future economically viable technology.

Examples

The literature contains many examples of TEAs of advanced biofuels processes that should be evaluated very carefully. Assumptions used in each study have a significant impact on the economic results. Relative uncertainty in estimated capital and operating costs can be quite high, and variations in reported scaling factors can lead to differences in estimated costs. Varying economic assumptions (debt/equity, debt interest rate and duration, target internal rate of return, plant life, cost basis year, and construction and start-up duration) can also lead to very different model economics.

A recent article compares the techno-economics of several thermochemical conversion routes to biofuels and bioproducts published in the literature from 2009 to 2015.[2] The TPIs range from \$217 million to \$752 million, and the MFSPs range from \$1.82/gal to \$7.29/gal. TEAs for biochemical conversion process are also prevalent in the literature. Estimates for bio-ethanol production from biomass have recently been evaluated,[3] and the relative techno-economics for bio-ethanol as a function of various pretreatment options can be found in the literature.[4, 5] The techno-economics of bio-butanol production as a function of biomass pretreatment technologies has also been recently reviewed.[6] Table 7.1 summarizes the TEAs for several biochemical and thermochemical conversion routes to advanced biofuels.

Table 7.1 Techno-Economic Analysis (TEA) Results for Several Biochemical and Thermochemical Conversion Routes.

Process	Product	Plant size (dry tons/day)	Plant life (years)	Capacity factor	Total capital investment ($MM)	Cost basis year	MFSP ($/gal)	Feedstock cost ($/BDT)	Debt/equity	Loan interest rate	Loan term (years)	References
Biochemical conversion												
Dilute acid/SSF	Ethanol	2205	20	96	376	2007	3.42	83	0/100			5
Dilute acid/SSF	Ethanol	2205	30	96	423	2007	2.15	58.5	60/40	8	10	7
Dilute acid/EH/bioconversion	Hydrocarbon (fatty acid intermediate)	2205	30	90	582.7	2011	5.35	80	60/40	8	10	8
Dilute acid/EH/catalytic conversion	Hydrocarbons (sugar intermediate)	2205	30	90	659.6	2011	4.05	80	60/40	8	10	9
DMR pretreatment/EH/anaerobic fermentation/upgrading	Diesel	2205	30	90	758	2016	2.49/2.47	71.26	60/40	8	10	10
Liquid hot water/EH/fermentation	Ethanol	2337	30	90	386.2	2014	6.74	178				4
AFEX/EH/fermentation	Ethanol	2337	30	90	415.9	2014	6.81	178				4
Thermochemical conversion												
In situ CFP/upgrading	Gasoline	2205	30	90	546	2011	3.46	80	60/40	8	10	11
	Gasoline	2205	30	90	590	2011	3.31	80	60/40	8	10	11

Continued

Table 7.1 Techno-Economic Analysis (TEA) Results for Several Biochemical and Thermochemical Conversion Routes.—cont'd

Process	Product	Plant size (dry tons/day)	Plant life (years)	Capacity factor	Total capital investment ($MM)	Cost basis year	MFSP ($/gal)	Feedstock cost ($/BDT)	Debt/equity	Loan interest rate	Loan term (years)	References
Ex situ CFP/upgrading												
Gasification/mixed alcohol	Ethanol	2205	20	96	214	2007	1.29	51	0/100			12
Gasification/MTG	Gasoline	2205	20	96	200	2007	1.95	51	0/100			13
High temp gasification/FTS	Gasoline/diesel	2205	20	85	657	2007	5.24	75				14
Low temp gasification/FTS	Gasoline/diesel	2205	20	85	540	2007	4.63	75				14
Fast pyrolysis/FTS	Diesel	2000	15	96	249	2011	3.3	61.2				15
Fast pyrolysis/upgrading/H_2 production	Gasoline	2205	20	96	287	2007	3.09	75	0/100			16
Fast pyrolysis/upgrading/H_2 purchase	Gasoline	2205	20	96	200	2007	2.11	75	0/100			16
Fast pyrolysis and hydrotreating bio-oil	Gasoline/diesel	2205	30	90	700.6	2011	3.34/3.71	80	60/40	8	10	17

SSF, simultaneous saccharification and fermentation; *DMR*, deacetylation and mechanical refining; *EH*, enzymatic hydrolysis; *AFEX*, ammonia fiber explosion; *CFP*, catalytic fast pyrolysis; *MTG*, methanol-to-gasoline; *FTS*, Fischer-Tropsch synthesis.

Close-up: Contributions of Techno-Economic Analysis to Advancing Bioenergy

Techno-economic analysis (TEA) is used by government, industry, and academia to quantify the impact of research discoveries and engineering advances on the economic viability of an integrated process. When effectively coupled with research and development, TEA is an important complementary tool for understanding and identifying key process attributes that can be improved to minimize production costs (Fig. 7.2). When performing TEA across a range of technology readiness levels (TRLs), different analysis strategies apply. For instance, early-stage TEA is used to validate an initial idea, while the development of more rigorous process designs and economic evaluation is done with high TRL technologies. This analysis is necessary to support research and development direction, identify production and cost drivers, and inform key decisions on cost-efficient strategies for bio-derived fuels and chemicals. Cost numbers are used to map and track individual program accomplishments to evaluate the potential for technical success; this allows inclusion of multiple products and processes in the aggregate net value estimate. Key outcomes of strategic TEA are not just cost numbers but are quantitative assessments and evaluations of a technology's cost

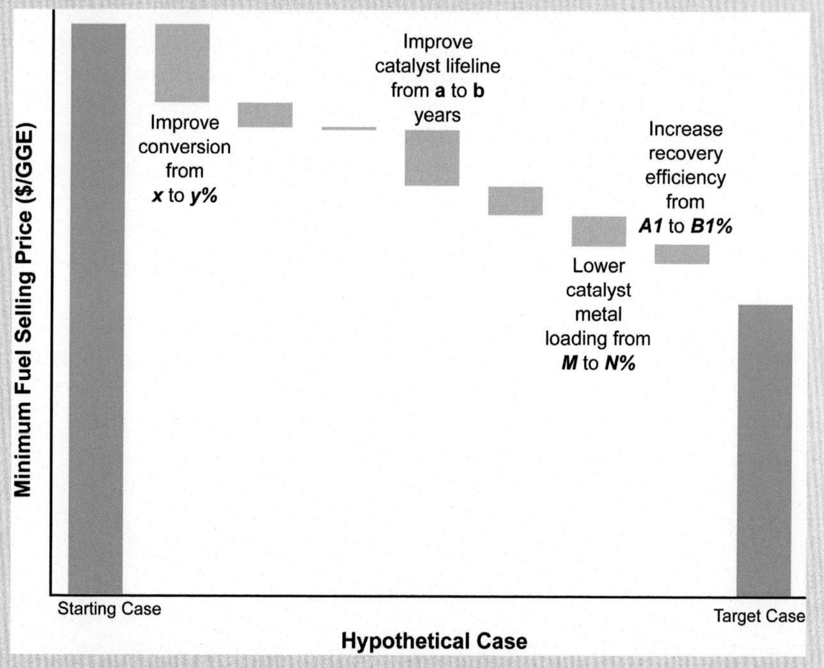

Fig. 7.2 Understanding the economic impact of overcoming technical barriers through targeted research and development.

Continued

Close-up: Contributions of Techno-Economic Analysis to Advancing Bioenergy—cont'd

reduction strategy, environmental impact, and scale-up challenges for the integrated advanced biorefinery of fuels and chemicals. Moreover, strategic TEA has been extended to complex sensitivity analysis, scenario analysis, comparative analysis, advance bioenergy concepts with multiple coproducts, biorefinery integration with existing supply chains, and infrastructures and biorefinery optimization.

Sensitivity analysis is performed to identify key cost drivers, and to address questions about economic viability and uncertainties, particularly for newly developed projects. Typical methods are single point sensitivity and multi-variant sensitivity, to which stochastic and Monte Carlo analyses are applied. *Scenario analysis* predicts or projects overall economic viability of a technology concept along its development timeline based on the level of investment and rate of technical progress. The economic viability, energy efficiency, and sustainability of a given process, along with risk mitigation drivers, can be calculated for several different scenarios, such as "business as usual," "realistic or state-of-technology," and "pessimistic" or "optimistic or future" cases.

Comparative TEA analysis is the best way to investigate a broad range of technologies for their economic and environmental sustainability potentials, to assist decision-making and strategic directions in advancing bioenergy. Economic evaluations compare technologies against a benchmark such as traditional refinery or fossil-based energy production, or compare alternative process strategies and chemistries to facility decision making. The key for a good comparative analysis is to set good reference points and a consistent basis to understand the tradeoffs and opportunities for alternative designs. The analysis includes evaluation of technologies with processing metrics (e.g., mass balance, carbon balance, energy efficiency, separation and reaction process synthesis, process complexity, cost potential, and TRL) and environmental or market related metrics (e.g., tradeoff strategies for business development, market size and saturation point, favorable life cycle inventory, industrial interests, and end uses).

Developing *analyses of integrated biorefineries* to evaluate the impact of products is an option to de-risk biorefinery economics. Utilization of co-products has supported corn ethanol and the petroleum refinering industry as an evolving optimization strategy. Maximizing facility profits could be the objective function of using the optimization tools in the petrochemical industry to evaluate a range of scenarios and market conditions for integrated biorefinery performance. Co-product accounting is necessary to take into account the benefits of fuels, electricity, and products from biomass. The responsibility of analysis is to develop methods for forecasting future

Close-up: Contributions of Techno-Economic Analysis to Advancing Bioenergy—cont'd

prices of commodity chemicals (including market considerations such as saturation). This would address the question of what price or range of prices should be used in cost evaluation for a chemical in an integrated biorefinery to support questions raised by the stakeholders. *Linear programing* (LP) *models* are developed to provide insightful and rapid analyses to guide advanced bioenergy research on the process development and valuation of products from an integrated biorefinery perspective, and to investigate the value and blend wall limits for blending biomass-derived biofuels with petroleum products. LP modeling is focused on the integration aspects of biofuel to existing energy infrastructure (such as aspects of technical feasibility), or on the assessment of the value proposition of biomass-derived blendstocks (such as aspects of economic feasibility) that may contribute to finished fuel blends in various petroleum refinery facilities. The goal is to investigate how, under variable economic situations, a biorefinery would react, and whether it would still be profitable in low crude price situations to have a diversified product portfolio.

Even more broadly, strategic TEA is coupled with supply chain analysis to explore potential for integrated landscape management and ecosystem services to reduce the cost of bioenergy, address strategic implications, measure impacts of bioproducts, and compile data across the supply chain. This facilitates global carbon accounting and the circular economy.

A mixed-integer LP model can be developed that takes into account the main characteristics of bioenergy supply chains, such as seasonality and regionality of feedstock supply, biomass deterioration with time, geographical diversity and availability of biomass resources, feedstock density, diverse conversion technologies and co-products, infrastructure compatibility, demand and market share distribution, regional economic structure, environmental and social impacts, and local or national government subsidies. Mixed-integer LP optimization integrates decision making across multiple temporal and spatial scales and simultaneously predicts the optimal network design, facility location, technology selection, capital investment, production operations, inventory control, and logistics management decisions. In addition to the economic objective of minimizing the annualized net present cost, the mixed-integer LP is integrated with life cycle assessment (LCA), feedstock resources and supply entry points through a multi-objective optimization scheme to include objectives on environmental and social metrics of the bioenergy or existing energy supply chain. The

Continued

Close-up: Contributions of Techno-Economic Analysis to Advancing Bioenergy—cont'd

effective linkage between TEA and LCA during an active research and development stage is of growing interest to the bioenergy community, with economic and environmental sustainability metrics being considered simultaneously to identify promising effects to advancing bioenergy.

Ling Tao

National Renewable Energy Laboratory, Golden, CO, United States

References

1. Aden A, Ruth M, Ibsen K, et al. *Lignocellulosic Biomass to Ethanol Process Design and Economics Utilizing Co-Current Dilute Acid Prehydrolysis and Enzymatic Hydrolysis for Corn Stover.* National Renewable Energy Laboratory; 2002.
2. Brown TR. A techno-economic review of thermochemical cellulosic biofuel pathways. *Bioresour Technol.* 2015;178:166–176.
3. Chovau S, Degrauwe D, Van der Bruggen B. Critical analysis of techno-economic estimates for the production cost of lignocellulosic bio-ethanol. *Renew Sustain Energy Rev.* 2013;26:307–321.
4. da Silva ARG, Torres Ortega CE, Rong B-G. Techno-economic analysis of different pretreatment processes for lignocellulosic-based bioethanol production. *Bioresour Technol.* 2016;218:561–570.
5. Kazi FK, Fortman JA, Anex RP, et al. Techno-economic comparison of process technologies for biochemical ethanol production from corn stover. *Fuel.* 2010;89:S20–S28.
6. Baral NR, Shah A. Comparative techno-economic analysis of steam explosion, dilute sulfuric acid, ammonia fiber explosion and biological pretreatments of corn stover. *Bioresour Technol.* 2017;232:331–343.
7. Humbird D, Davis R, Tao L, et al. *Process Design and Economics for Biochemical Conversion of Lignocellulosic Biomass to Ethanol: Dilute-Acid Pretreatment and Enzymatic Hydrolysis of Corn Stover.* Golden, CO: National Renewable Energy Lab. (NREL); 2011. NREL/TP-5100-47764; TRN: US201110%%638 United States 10.2172/1013269 TRN: US201110%%638 NREL English.
8. Davis R, Tao L, Tan ECD, et al. *Process design and economics for the conversion of lignocellulosic biomass to hydrocarbons: dilute-acid and enzymatic deconstruction of biomass to sugars and biological conversion of sugars to hydrocarbons.* October 2013. NREL/TP-5100-60223.
9. Davis R, Tao L, Scarlata C, et al. *Process design and economics for the conversion of lignocellulosic biomass to hydrocarbons: dilute-acid and enzymatic deconstruction of biomass to sugars and catalytic conversion of sugars to hydrocarbons.* March 2015. NREL/TP-5100-62498.
10. Davis RE, Grundl NJ, Tao L, et al. *Process design and economics for the conversion of lignocellulosic biomass to hydrocarbon fuels and coproducts: 2018 biochemical design case update; biochemical deconstruction and conversion of biomass to fuels and products via integrated biorefinery pathways.* 11/19/2018. NREL/TP-5100-71949.
11. Dutta A, Sahir A, Tan ECD, et al. *Process design and economics for the conversion of lignocellulosic biomass to hydrocarbon fuels thermochemical research pathways with in situ and ex situ upgrading of fast pyrolysis vapors.* March 2015. NREL/TP-5100-62455.
12. Dutta A, Phillips SD. *Thermochemical ethanol via direct gasification and mixed alcohol synthesis of lignocellulosic biomass.* NREL/TP-510-45913.

13. Phillips SD, Tarud JK, Biddy MJ, Dutta A. Gasoline from Wood via Integrated Gasification, Synthesis, and Methanol-to-Gasoline Technologies. *NREL Publications Database*. 2011. 115 NREL/TP-5100-47594. https://www.nrel.gov/docs/fy11osti/47594.pdf.
14. Swanson RM, Platon A, Satrio JA, Brown RC. Techno-economic analysis of biomass-to-liquids production based on gasification. *Fuel*. 2010;89:S11–S19.
15. Manganaro JL, Lawal A. Economics of thermochemical conversion of crop residue to liquid transportation fuel. *Energy Fuel*. 2012;26(4):2442–2453.
16. Wright MM, Satrio JA, Brown RC, Daugaard DE, Hsu DD. *Techno-economic analysis of biomass fast pyrolysis to transportation fuels*. November 2010. NREL/TP-6A20-46586.
17. Jones S, Meyer P, Snowden-Swan L, et al. *Process design and economics for the conversion of lignocellulosic biomass to hydrocarbon fuels: fast pyrolysis and hydrotreating bio-oil pathway*. November 2013. PNNL-23053.

Additional Reading

Peters M, Timmerhaus K, West R, Peters M. *Plant Design and Economics for Chemical Engineers*. 5th ed. McGraw-Hill Education; 2002.

CHAPTER EIGHT

Lifecycle Assessment of Advanced Biofuels Processes

Introduction

"Paper or plastic?" A simple enough question; but what factors go into determining which is the better option? Cost, convenience, and consumer preference are important. For some consumers, environmental performance and impact are important considerations for product selection, but the relative environmental performance of specific options is difficult to understand and quantify. More frequently, manufacturers of products are looking for "greener" options to meet consumer expectations and to become more sustainable. These companies need a way to quantify environmental impacts to justify and defend their choices for improving sustainability. Lifecycle assessment (LCA) is a tool used to inform decisions about the environmental performance and impact of products and processes.

LCA is a comprehensive and transparent method to account for the environmental (ecological and health) impacts of a product or process "from cradle to grave." Defining what that means for a specific system is important for analyzing environmental trade-offs between different products and assuring that impacts are not overlooked because they are shifted to a different, potentially unidentified lifecycle stage or effluent.

For this discussion, the product is renewable fuel that can be produced by several different processes, and the system boundary encompasses the entire value chain from producing and harvesting biomass through biofuel use. Renewable fuels produced from biomass feedstocks are expected to have lower greenhouse gas (GHG) emissions than petroleum-based transportation fuels and, consequently, less impact on global climate change. One of the objectives of LCA is to validate the GHG emissions reduction potential of renewable fuel options. Global climate change has been a strong driver for renewable energy policy, and LCA has become the standard for

Analytical Methods for Biomass Characterization and Conversion
https://doi.org/10.1016/B978-0-12-815605-6.00008-1

quantifying and comparing GHG emissions from the many commercial and developing biofuels technology options.

The Renewable Fuel Standard (RFS) was established as part of the Energy Policy Act of 2005 and expanded under the Energy Independence and Security Act of 2007 to provide a market pull for developing technologies to increase production of advanced biofuels. The goal of the RFS was to reduce the impact of transportation fuels on global climate change, improve energy security in the United States, and increase rural economic development through the production and use of renewable fuels.

The RFS program is a national policy directed by the Administrator of the United States Environmental Protection Agency (EPA) to reduce the quantity of fossil fuels used for transportation by establishing annual volume targets for renewable fuel production between 2006 and 2022. Renewable fuel is defined as transportation fuel produced from renewable biomass that displaces fossil fuel. Renewable fuels include conventional biofuel, biomass-based diesel, advanced biofuels, and cellulosic biofuel. The term advanced biofuel is defined as a renewable fuel, other than ethanol produced from corn starch, that has at least 50% lower GHG emissions compared to a baseline established for petroleum-derived gasoline and diesel transportation fuels. Cellulosic biofuels are produced from cellulose, hemicellulose, or lignin, with at least a 60% lifecycle GHG emissions reduction compared to fossil fuels. Examples of advanced biofuels are cellulosic ethanol, bio-butanol, biomass-based diesel, cellulosic biofuel, and bio-gas.

Specific batches of renewable fuel are assigned Renewable Identification Numbers (RINs) to track production, use, and trading as part of the RFS. RINs are generated by renewable fuel producers, traded by transportation fuel market participants (refiners, distributors, blenders, and wholesalers), and ultimately retired by obligated parties for compliance to the RFS. RIN values are established in accordance with the type of fuel and are effectively the currency of the RFS. Each fuel is assigned a D-code that identifies the type based on the feedstock used to produce it, the type of fuel produced, and the greenhouse emission reduction potential. Table 8.1 lists the RIN codes for five fuel types.

Lifecycle GHG emissions refers to the direct and indirect release of CO_2, methane, nitrous oxide, halocarbons, and sulfur hexafluoride during all stages of feedstock production and collection, fuel production, distribution, and delivery to the end-user. All GHG emissions are typically normalized to CO_2 equivalents based on their relative global warming potential. Lifecycle GHG emissions are the basis for establishing renewable energy policy, determining compliance with legislation, and determining the value of RIN credits.

Table 8.1 RIN D-Codes for Five Types of Renewable Fuel.

Fuel	Feedstock	GHG emissions reduction (%)	RIN D-code
Cellulosic Biofuel	Agricultural crops and residues, trees and tree residues, animal wastes and byproducts, forest thinnings, algae, yard waste, food waste (including waste grease and oils)	60	D3
Cellulosic Diesel		60	D7
Advanced Biofuel		50	D5
Biomass-Based Diesel	Oil seed crops, vegetable oils	50	D4
Conventional Renewable Fuel	Corn starch	20	D6

GHG emissions from advanced biofuels technology pathways cannot be measured directly, because they are still under development. Therefore, the method for determining GHG emissions reduction potential for developing technologies includes estimates based on the rigorous process models and material and energy balances that were the basis of the techno-economic analyses discussed in Chapter 7. Determining the lifecycle GHG emissions for specific renewable fuel production pathways becomes important to determine which RIN credit, if any, a renewable fuel qualifies for.

The methodology for determining lifecycle GHG emissions of transportation fuels has received intense scrutiny because of their political and economic significance. The system definition and boundaries are critical for assessing the sustainability of renewable fuels. But setting system boundaries can be challenging, because they can have a direct influence on the interpretation of LCA results. Selecting system boundaries should be consistent to facilitate comparison of the relative sustainability of different biofuel pathways. The boundaries should also be selected so that enough scientific rigor can be incorporated to defend the LCA.

Methodology

The International Organization for Standardization has developed the standard (ISO 14040:2006) for an internationally recognized methodology that defines the principles and framework of LCA. It is an iterative process that starts by defining the goal and scope of the analysis and proceeds with developing an inventory of all inputs and outputs to the system and assessing the environmental impact, while constantly interpreting the data and the results and continually updating and improving each step of the process.

For this discussion, the product to be evaluated is renewable fuel. The LCA for renewable fuels accounts for GHG emissions from (1) biomass production, harvesting, and processing, (2) biomass conversion and upgrading to biofuel, (3) biofuel use for transportation, (4) end of life disposal of equipment and facilities, and (5) transportation and distribution between and within the boundary of the defined system and subsystems. Fig. 8.1 depicts the framework of a typical biofuels LCA. The goal is to compare the GHG emissions of developing advanced biofuels with conventional petroleum-based transportation fuels. Establishing the system boundaries for the baseline (petroleum fuels) and advanced biofuels sets the scope of the comparison and defines the inputs and outputs that need to be inventoried. The impact of renewable fuels development and use on global climate change is the primary driver; however, water usage, air emissions (other than GHG), and indirect land use changes can also be assessed.

Two modeling approaches are available for LCA, depending on the stated goal of the assessment. An attributional LCA defines allocations of environmental impact to the physical flows within the system over the stated lifecycle time frame (the life of the biofuels plant, for example). Average emissions over the lifecycle are used for energy inputs and outputs. An **attributional** LCA can be used to compare the relative environmental impacts

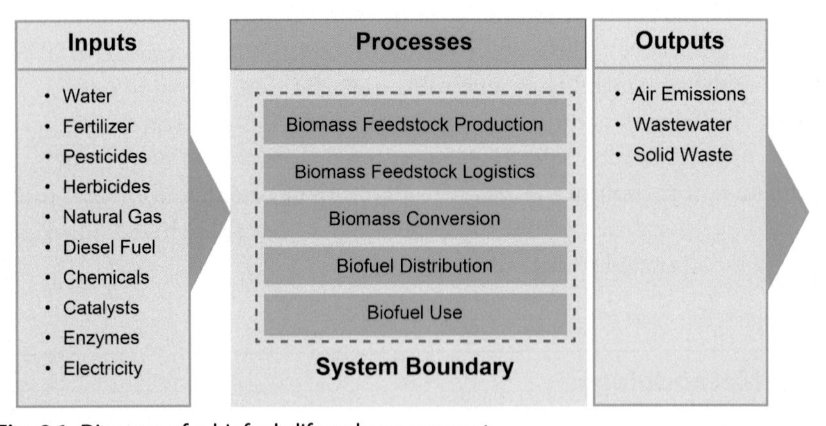

Fig. 8.1 Diagram of a biofuels lifecycle assessment.

of different products or different processes. A consequential LCA, in contrast, looks at the long-term impacts of the system beyond the immediate changes allocated to the system. A **consequential** LCA seeks to assess the future impacts of decisions. For example, as biofuels capture a larger share of the transportation market, demand for petroleum may decrease (reducing fossil GHG emissions), but increased biofuels production will lead to higher biomass demand and potentially accelerate the impact of indirect land use.

Defining a functional unit for the LCA dictates how the data will be collected and organized. For biofuels LCA there are two common options to assess GHG emissions reduction. Well-to-tank analysis defines the system boundary to encompass the relative GHG emissions for fuel production. The functional unit is reported as GHG emissions per gallon of fuel produced or energy content of the fuel. This type of analysis is useful for assessing the relative carbon and energy efficiency for specific conversion technologies and any differences in feedstock production and delivery. Well-to-wheels analysis expands on well-to-tank to include biofuels utilization. The functional unit is GHG emissions per vehicle mile traveled. Engine performance and fuel efficiency are accounted for in well-to-wheels. For example, diesel fuel has lower GHG emissions per vehicle mile than gasoline, because diesel engines (compression ignition) are more efficient than gasoline engines (spark ignition). Also, the energy density of ethanol is two-thirds that of gasoline. Therefore, in a standard internal combustion engine, fuel efficiency decreases as a function of the percentage of ethanol in the blend. The amount of biofuel being used is increasing but the number of miles traveled per gallon of fuel is decreasing. On the other hand, ethanol has a higher octane rating than gasoline. High compression ratio engines can take advantage of this higher octane for improved efficiency and reduce the number of miles traveled per gallon of fuel.

Lifecycle Inventory

The lifecycle inventory compiles all energy and raw material inputs and air emissions, wastewater, and solid waste outputs in terms of the functional unit defined for the assessment. For biofuels processes, a distinction is made between biogenic and fossil carbon. Biogenic carbon originates from a biological source and can be released to the atmosphere as CO_2 during biofuel use or biomass combustion and removed from the atmosphere through photosynthesis for biomass production. Therefore, there are no net CO_2

emissions from biogenic carbon in a biofuels process for the purpose of LCA. Fossil carbon originates from fossil fuel reserves accumulated over a geologic timescale and can only be released to the atmosphere when consumed. GHG emissions for each process, shown in Fig. 8.1, are compiled and defined as biogenic or fossil carbon.

LCA and emissions data are gathered from many sources. It can be measured directly or compiled from a variety of literature resources. Fortunately, there are also many LCA modeling packages that contain comprehensive databases. The data is also available from the heat and material balances calculated from the detailed process models for specific biofuels technologies.

Biomass Feedstock Production

The lifecycle inventory for biomass production is strongly dependent on the type of biomass and includes estimates of above- and belowground carbon and impacts of direct and indirect land use changes. For example, inventories for agricultural residues such as corn stover and wheat straw need to consider farming practices. GHG emissions associated with, field preparation (tilling vs no-till, herbicide production and application, planting), fertilizer production and application, irrigation, crop harvesting, and residue collection all need to be included in the inventory but vary during the growing season. One challenge is allocating GHG emissions in the LCA between the food crop and the residue used for biofuels. Another is estimating the amount of carbon sequestered in the soil compared to the amount of crop residue that is removed from the field. Soil carbon includes the root mass of the crops and the amount of crop residue that is plowed back into the ground. CO_2 emissions from farming equipment powered by diesel fuel also need to be accounted for, as do nutrient runoff and post-harvest erosion.

Food crops are annuals that require intensive farming each year. Energy grasses like switchgrass and miscanthus are perennials, so the field maintenance intensity varies. Initiation may require intensive field preparation and fertilizer application, but these should lessen over time as the grass is established. It may be possible to harvest energy grasses multiple times a year, affecting the GHG emissions for feedstock collection. Another impact of energy grasses is associated with land use change. Often, marginal lands and in some cases pasture lands are considered for energy grass production. The environmental impacts can shift as these low-maintenance lands are put into active biomass production.

Lifecycle environmental impacts of woody biomass collection are strongly dependent on forest management practices. Like food crops, trees have multiple uses, such as lumber, wood products, pulp and paper, and energy. Forest thinnings are targeted for biofuels production, so lifecycle environmental impacts need to be allocated between the various products and biofuels. Tree plantations require initial field preparation for planting saplings and establishing stands. These stands may be initially irrigated and managed, but less maintenance is needed as the trees mature. Whereas crops and energy grasses can be harvested annually or more often, trees are harvested on a much longer time scale that may exceed that of the LCA. Some fast-growing trees are harvested every 5–7 years; softwoods for lumber may be harvested every 30–35 years. Therefore, to sustainably harvest trees, a large land radius is required to accommodate the growing cycle. Clear-cutting forests may initially provide a windfall of biomass, but the long-term environmental impacts can be severe. The amount of CO_2 taken up by the trees is also a function of the age of the trees, the average climate, and the season. The amount of CO_2 emissions from forestry equipment must also be accounted for in the lifecycle inventory.

Biomass Feedstock Logistics

Lifecycle impacts of biomass feedstock logistics are associated with processing and delivering feedstock to the conversion plant gate. For agricultural residue and energy grass, GHG emissions from equipment for harvesting, chopping, and baling are counted. Harvesting, chipping, and grinding equipment for forest residue is also included. GHG emissions from diesel-powered trucks to transport the bales or chips to the plant or a storage facility are accounted for. In some locations, feedstock can also be transported by rail or barge.

Biomass Conversion

GHG emissions from conversion are estimated from the detailed process designs for developing technologies. For biochemical conversion, lifecycle impacts of enzyme production and fermentation are included. Heat and electricity requirements can be partly or totally met within the process. Lignocellulosic ethanol process designs often include a power generation section that produces excess electricity from lignin combustion. This green power can be exported to the grid for a GHG emissions credit. For thermochemical processes, heat integration reduces the need for supplemental energy; however, lifecycle impacts of catalyst production and hydrogen production (from natural gas reforming) for upgrading can include fossil fuel input. GHG emissions credits can be realized if excess char is recovered and used (sequestered) as a soil amendment or burned for green power production.

Biofuel Distribution

Biofuel (ethanol, biobutanol, or hydrocarbons) distribution includes transport to a terminal or blending facility and distribution of the blended fuel to the end-user. It can be achieved by pipeline, truck, rail, or barge, depending on the biofuel and the proposed location of the conversion facility. Environmental impacts are allocated between the different transportation methods. In addition, how these modes of transportation are fueled can affect the LCA, especially if 100% biofuel instead of petroleum-derived fuel is used.

Biofuel Use

Engine performance and fuel efficiency are factored into the GHG emissions per vehicle mile traveled in a well-to-wheels analysis. Diesel engines are more efficient than gasoline engines, primarily because diesel has a higher energy density. Biodiesel produced from transesterification of vegetable oils is blended with petroleum diesel in 5% and 20% ratios. Renewable diesel from hydroprocessed vegetable oils and drop-in hydrocarbons from advanced biofuels processes typically don't have a blend limit. Conventional gasoline contains 10% ethanol, but higher percentages are available (E15 and E85). Unless the engine is optimized for the increased octane of higher ethanol blends, fuel efficiency decreases.

Tailpipe emissions other than CO_2 are usually not accounted for in LCA, because the fuels and fuel blends have been certified by the EPA. In the past, soot from diesel engines and increased volatile organic hydrocarbon emissions from gasoline engines running on high ethanol blends were a concern, but electronic fuel injection in newer vehicles minimizes those emissions.

Environmental Impact Assessment

A comprehensive lifecycle inventory can be analyzed to link a specific biofuels process with environmental impacts associated with resource depletion and air, water, and solid waste emissions. The goal of the environmental impact assessment is not to quantify the environmental effects of the product or process but to understand the link between stages of the LCA or aspects of the process to specific environmental effects. After identifying the relevant impacts and assigning the results from the lifecycle inventory to specific impacts, a sensitivity analysis reveals the input variables that have the largest effect on the output emissions and thus on the environment.

GHG emissions reduction by displacing fossil fuels with biofuels is the primary focus of biofuels LCA for compliance with biofuels policy to mitigate global warming. All air emissions from the entire value chain for biofuel production and use are normalized to CO_2 equivalents based on relative global warming potential and compared to fossil-based transportation fuel emissions. Besides global warming, air emissions can have other negative environmental effects. Nitrogen oxide contributes to smog formation that reduces air quality. Sulfur dioxide contributes to acid rain, which leads to deforestation. Chlorofluorocarbon and hydroflurocarbon emissions are responsible for stratospheric ozone depletion.

The environmental impacts of water use, treatment, and discharge are becoming increasingly important for the use of biomass in fuel production. Water runoff from cultivated farmland can carry excess fertilizer to streams and rivers, causing watershed eutrophication. An extreme example is the "dead zone" in the Gulf of Mexico, created by fertilizer runoff carried by the Mississippi River; local problems can also arise. Biofuel LCAs are also being used to determine the environmental impact on local and regional water cycles. Fresh or reclaimed water is needed for biomass production, and water use and reuse in the conversion process require careful management. Water use effects can be a resource depletion issue in arid regions and a discharge issue in locations with inadequate infrastructure to manage wastewater treatment.

The environmental impact of direct and indirect land use changes induced by biofuels production has drawn a lot of recent attention. Concerns are split between increasing the intensity of how existing croplands and forests are managed to meet the needs of the developing biofuels industry, and transforming idle lands or less intensely managed lands into biomass production for biofuels. Land resources have developed over time to meet the societal demands for food, feed, and fiber. Additional biomass production cannot be allowed to disrupt current land use requirements, bringing to bear the question of biomass sustainability. Direct land use changes are often understood, but uncertainty comes from estimating the lifecycle impacts associated with indirect land use changes.

A well-known outcome of indirect land use change is the "food vs fuel" debate. The argument was that increasing corn ethanol production in the mid-2000s was driving up the price of food; corn is used to feed livestock and humans, but more corn was going into ethanol production for transportation. This argument has since been disproved, but it is one of the rationales for advanced biofuels technologies that use lignocellulosic feedstocks instead of food grains. Another unintended consequence that is being studied is loss of biodiversity as birds and small animals lose habitat.

Bioenergy crops, however, can have benefits when planted alongside food crops or in managed forests. Bioenergy crops in buffer regions or riparian zones help manage nutrient runoff. They can also be planted on idle or marginal land to control erosion and increase soil carbon. These benefits are known as ecosystem services, but the environmental impacts are difficult to quantify in an LCA.

Conclusion and Remaining Challenges

The LCA is the basis for establishing renewable energy policy, determining compliance with legislation, and determining the value of renewable fuel credits. Ambitious targets have been established to increase biofuels production to slow the impact of transportation on global climate change. At the same time, the environmental sustainability of biofuels production was being questioned. Does the production and use of advanced biofuels actually have lower GHG emissions than fossil fuels, when all else is considered?

GHG emissions from biofuel use are assumed to be zero because there are no fossil emissions associated with the components in the system that originate from biomass. But if biomass is not harvested for biofuel production, it can still grow and take up atmospheric CO_2. To be sustainable, biomass resources need to be replenished to capture CO_2. But when biomass is harvested, carbon remains sequestered in the soil if it is not disturbed. Consequently, the impact of indirect land use changes continues to be studied in relation to sustainable biomass production to reduce the uncertainty of unintended consequences in LCA. This highlights the importance of LCA to untangle the connections in these complex systems and thereby assess the environmental impact of new and developing biofuels technology pathways.

The initial focus was on biofuels for road travel, but recently there has been a push to develop alternative jet fuel. The Commercial Alternative Aviation Fuels Initiative (CAAFI; see www.caafi.org) is a coalition of airlines, original equipment manufacturers, alternative fuel developers, and researchers, formed to promote alternative options for jet fuel that reduce price volatility and decrease GHG emissions by displacing fossil fuels without compromising safety or performance. Fuel certification and qualification assure that alternative fuels can meet the rigorous standards for aviation. LCA is the basis for defining the sustainability of approved alternative jet fuel pathways. Before a new pathway is approved, fuel certification standards must be met, and the GHG emissions reduction potential of the new pathway must equal or exceed 50% compared to petroleum-based jet fuel.

The environmental and social sustainability of biofuels continues to be scrutinized, and LCA methodology is constantly evolving. Access to quality data is a must to refine assumptions and reduce uncertainty to allow better assessment of the environmental impacts of complex, interrelated systems for biofuels production and use. Numerous LCAs have recently been published addressing the environmental sustainability of various biofuels technology pathways.[1–10] The Additional Reading list gives more general resources on LCA methodology and best practices.

References

1. Borrion AL, McManus MC, Hammond GP. Environmental life cycle assessment of lignocellulosic conversion to ethanol: a review. *Renew Sust Energy Rev*. 2012;16(7): 4638–4650.
2. Hsu DD, Inman D, Heath GA, Wolfrum EJ, Mann MK, Aden A. Life cycle environmental impacts of selected U.S. ethanol production and use pathways in 2022. *Environ Sci Technol*. 2010;44(13):5289–5297.
3. Gnansounou E, Dauriat A, Villegas J, Panichelli L. Life cycle assessment of biofuels: energy and greenhouse gas balances. *Bioresour Technol*. 2009;100(21):4919–4930.
4. Hsu DD. Life cycle assessment of gasoline and diesel produced via fast pyrolysis and hydroprocessing. *Biomass Bioenergy*. 2012;45:41–47.
5. Kendall A, Yuan J. Comparing life cycle assessments of different biofuel options. *Curr Opin Chem Biol*. 2013;17(3):439–443.
6. Malca J, Coelho A, Freire F. Environmental life-cycle assessment of rapeseed-based biodiesel: alternative cultivation systems and locations. *Appl Energy*. 2014;114:837–844.
7. McKone TE, Nazaroff WW, Berck P, et al. Grand challenges for life-cycle assessment of biofuels. *Environ Sci Technol*. 2011;45(5):1751–1756.
8. Wang M, Han J, Dunn JB, Cai H, Elgowainy A. Well-to-wheels energy use and greenhouse gas emissions of ethanol from corn, sugarcane and cellulosic biomass for US use. *Environ Res Lett*. 2012;7(4):1–13.
9. Davis SC, Anderson-Teixeira KJ, DeLucia EH. Life-cycle analysis and the ecology of biofuels. *Trends Plant Sci*. 2009;14(3):140–146.
10. Guinée JB, Heijungs R, Huppes G, et al. Life cycle assessment: past, present, and future. *Environ Sci Technol*. 2011;45(1):90–96.

Additional Reading

Curran MA. Life cycle assessment: a review of the methodology and its application to sustainability. *Curr Opin Chem Eng*. 2013;2(3):273–277.

Curran MA. *Lifecycle Assessment: Principles and Practice*. U.S. Environmental Protection Agency; 2006. EPA/600/R-06/060.

Hauschild MZ, Rosenbaum RK, Olsen SI. *Life Cycle Assessment: Theory and Practice*. Springer International Publishing; 2017.

PUBLIC LAW 110-140. *Energy Independence and Security Act of 2007*. US Government Publishing Office; 2007:341.

CHAPTER NINE

Optimized Biofuels for High-Efficiency, Low-Emission Engines

Introduction

The overwhelming majority of fuels are used in vehicles for the transport of people and goods. The three main fuel types are gasoline for spark ignition (SI) engines in passenger vehicles and light trucks, diesel for compression ignition (CI) engines in medium and heavy-duty transport, and jet fuel for jet powered air transport. Although an exact number is difficult to establish, current estimates indicate that there are approximately 1.4 billion cars, trucks, buses, and airplanes in service worldwide, with a doubling rate of approximately 20 years. The impacts of all these vehicles touches many elements of human life in both positive and nonpositive ways and will continue to do so for the foreseeable future. Many of the nonpositive impacts center around the emissions and environmental impacts of all these vehicles, so continuing to improve the performance and efficiency and reduce the environmental impact is a global issue of utmost importance and urgency.

A major driver for improved engine technologies has long been the increasingly stringent exhaust emission regulations that are necessary to maintain air quality standards in the face of increasing vehicle use. The key emissions of concern are oxides of nitrogen (NO_x) and sulfur (SO_x), particulate matter (PM or soot), unburned hydrocarbons (UHCs), and associated oxygenates such as aldehydes, carbon monoxide, and carbon dioxide. Nitrogen dioxide readily forms during the combustion process from nitric oxide in the atmosphere. It has many detrimental effects on human health and air quality. It reacts with ammonia and moisture to create small particles that cause or aggravate respiratory and heart disease. It also reacts with volatile organic compounds such as UHCs to form ground-level ozone. Ozone is a major contributor to urban smog and respiratory problems, especially in people with asthma. NO_x also reacts with oxygen, water, and other

Analytical Methods for Biomass Characterization and Conversion
https://doi.org/10.1016/B978-0-12-815605-6.00009-3

contaminants in the atmosphere to form nitric acid, which causes acid rain, which leads to acidification of lakes and streams, lowering the pH and releasing toxic aluminum that harms aquatic life. Acid rain produces slower growth or death of vegetation and forests and erodes infrastructure such as bridges and buildings. SO_x also causes respiratory problems and contribute to acid rain and urban smog; fuel based sulfur is the predominant contributor. Standards introduced in 2007 to reduce sulfur in diesel to <15 ppm and in 2017 to reduce sulfur in gasoline to <10 ppm on a yearly averaged basis have significantly reduced SO_x emissions.

Motor vehicle PM emission is a major direct contributor to urban and global air pollution and negatively affects climate, environment, and public health.[1] It is among the top 10 causes of morbidity and mortality and is estimated to cause 6 million premature deaths annually.[2] Particulates are especially harmful when they are smaller than 10 μm in size, because these are transported deeper into the lungs and do the most harm to the respiratory system. PM also causes haze and esthetic damage to buildings and stone infrastructure, such as making statues dark and dingy.

Oxygenated UHCs such as aldehydes and ketones are currently not regulated but are considered toxic. Carbon monoxide is very detrimental to health because it reduces the oxygen-carrying capacity of the blood. It and can cause extreme respiratory distress and death at high concentrations. Carbon dioxide emissions depend upon the power produced, engine efficiency, and the hydrogen-to-carbon ratio of the fuel.

Great strides have been made in improving engines. Current vehicles have superior fuel efficiency, greater power, and much lower emissions than earlier generations. According to the United States Environmental Protection Agency,[3] a 2015 model passenger vehicle is roughly 99% cleaner for common pollutants (UHCs, carbon monoxide, NO_x, and PM) than a 1970 model. This is also true for trucks and buses. However, with worldwide vehicle use and distance traveled increasing, vehicles will need to be ever cleaner.

As engines become more efficient and less polluting, fuel properties become more important. Manufactures are improving efficiency by increasing compression ratio, using direct injection, and downspeeding in SI engines, and by decreasing compression ratio and downspeeding in CI engines. Manipulation of the Research Octane Number (RON), sensitivity, and charge cooling for SI fuels and cetane, volatility, and viscosity for CI fuels, is critical to enabling these efficiency advances while reducing emissions. Increased efficiency and reduced emissions do not necessarily go hand in hand; for instance, direct injection of fuel into the cylinder of an SI engine increases efficiency but leads to higher PM emissions under some operating conditions such as cold start and heavy load.

Biofuels rationally developed with the desired chemical and physical properties can provide dual benefits to increasing efficiency and reducing emissions. The development of ethanol and biodiesel as biofuels was mostly motivated by what could be cost competitive in the near term. Then their limitations, such as blending limits and existing fuel distribution infrastructure, led to a focus on gasoline, diesel, and jet hydrocarbons that could be produced directly from biomass. However, neither approach really considered how to advantageously use biofuel properties as an enabler for higher-efficiency, lower-emission engines. This chapter addresses the issue by examining the future fuel needs for each of the three main engine types and the fuel analysis tools required.

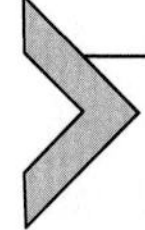

Spark Ignition Engines

History and Use

Many scientists and engineers contributed to the development of the modern SI engine, with early work going back to the late 18th century. In 1794, Robert Street built an internal combustion engine that used liquid fuel as its energy source. Throughout the 19th century, patents were issued for improvements on this original design to increase power, decrease weight, and enhance utility. Karl Benz invented the automobile in 1886. The introduction of the Ford Model T in 1908 made it affordable and created a mass market. The deployment of vehicles since then has been astounding, with an estimated 1.4 billion in service in 2015, of which 950 million were passenger cars powered mostly by SI engines. The positive impacts on society are hard to overstate: These vehicles enable fast, low cost transport of people and goods over long distances, which was not previously possible.

Throughout most of the history of passenger cars, the SI engine was the predominant power choice, with the CI engine an alternative. CI cars are more common in European markets. With advances in battery technology, electric vehicles are now commercially viable; and although they still have only a very small part of the market, rapid growth is projected for coming decades. However, the SI engine is likely to remain a major power source for passenger cars for decades because of low cost, ease of manufacture, and consumer acceptance.

Operation and Fuel Needs

Fig. 9.1 is a schematic of the SI combustion process for direct injection (left) and traditional port fuel injection (right). In port fuel injection, the gasoline is premixed with combustion air in the intake manifold upstream; then the vaporized, well mixed fuel-air charge is drawn into the cylinder through the intake valve. In direct injection, the gasoline is sprayed directly into the cylinder, which avoids the pumping losses associated with port fuel injection, but the fuel-air mixture is not premixed, which can increase PM emissions. A spark ignites the mixture based on engine timing within the piston's cycle of motion, and this spark-ignited flame kernel develops into a high-temperature turbulent premixed flame that propagates through the fuel-air charge. The fuel to air ratio in an SI engine must be kept very close to the stoichiometric mixture needed for complete combustion in order to use a three-way catalytic converter, which is very effective at reducing carbon monoxide, NO_x, and UHC emissions.

Important parameters for a gasoline fuel are octane rating, which is a measure of its autoignition properties (the ability of the fuel to prevent engine knock), turbulent flame speed, and emissions. Knock is a significant concern and occurs when a pocket of unburned fuel-air mixture spontaneously ignites and results in a rapid pressure rise in the cylinder. Severe knock can damage the engine. Higher compression ratios increase efficiency but also increase susceptibility to knock, so they require higher octane fuel.

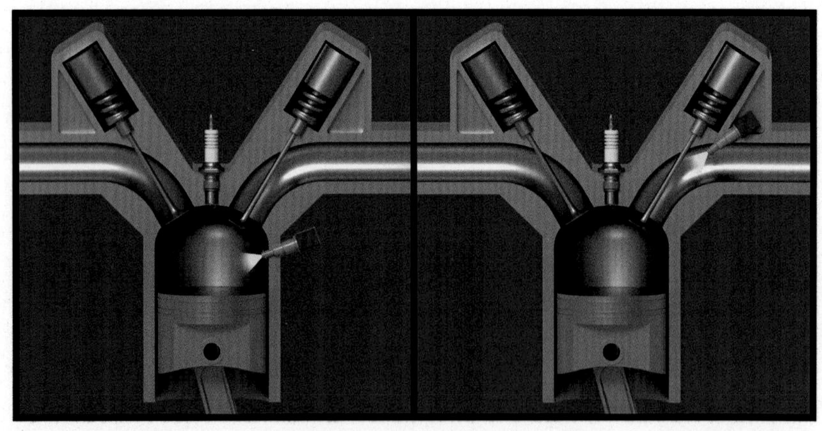

Fig. 9.1 Combustion process of the spark ignition engine for direct injection (left) and traditional port fuel injection (right).

Differences in octane ratings are related to differences in the fuel's low-temperature ignition chemistry, with low octane being readily ignitable and high octane more difficult to ignite. Autoignition is a complex process directly related to the propensity of the fuel to undergo ignition facilitated by the low-temperature "cool flame" chemical characteristics. The propensity to knock is directly related to the ability to undergo cool flame reactions to enable autoignition at lower temperature. Higher octane fuels, such as highly branched alkanes, are less reactive at low temperatures than straight chain hydrocarbons, which exhibit higher levels of low-temperature chemistry and thus have lower octane numbers. Generally, the tendency for knock increases with hydrocarbon chain length in alkane fuels, and aromatic compounds have very low knock potential.

Octane measurement uses Research Octane Number (RON) and Motor Octane Number (MON). Both are measured in a test engine according to a scale where *n*-heptane is defined as zero and isooctane as 100 for RON. Pump octane or anti-knock index is (RON + MON)/2. The primary differences between the two measurements are the fuel-air charge temperature and the engine speed. RON is done at a comparatively low fuel-air charge temperature and slower engine speed, MON at a higher fuel-air charge temperature and faster engine speed. Sensitivity (S) is the difference between RON and MON (S = RON − MON) and is considered a good metric for modern engines, where higher S is desired and can be achieved by increasing RON or lowering MON or both.

Certain biofuels, specifically alcohols, have very high RON values and good gasoline fuel properties in general. Table 9.1 lists methanol through pentanol properties in comparison with gasoline. Interesting trade-offs occur across the alcohols. For the lower ones (methanol and ethanol), RONs are high (110 and 109 respectively) but with low heating values (21.26 and 28 MJ/kg). For the higher ones, heating values are higher (33.1 and 36.4 MJ/kg for 1-butanol and 2-methyl-2-pentanol) but with lower RONs (98 and 99.2).

Another attractive property of biofuels is their low sooting propensity, measured by the yield sooting index (YSI). The higher the number, the greater the propensity to soot.[4] This is a reference scale defined as YSI = 0 for hexane and YSI = 100 for benzene, so negative values indicate lower sooting than hexane. All of these alcohols have low YSI values compared to gasoline, indicating that blending into gasoline will reduce soot emissions, which is very attractive for SI fuel blending components.

Table 9.1 Fuel Properties of Alcohols Compared to Gasoline.

Property	Gasoline	Methanol	Ethanol	1-Propanol	1-Butanol	2-Methyl-1-propanol	2-Methyl-2-pentanol	3-Methyl-3-pentanol
Molecular weight (g/mol)	100	32.04	46.07	60.1	74.12	74.12	102.17	102.17
C (wt%)	87	37.48	52.14	59.96	64.82	64.82	70.53	70.53
H (wt%)	13	12.58	13.13	13.42	13.60	13.60	13.81	13.81
O (wt%)	0	49.93	34.73	26.62	21.59	21.59	15.66	15.66
Boiling point (°C)	–	64.7	78.2	97.2	117.7	107.8	121.1	122.40
Melting point (°C)	–	−97.6	−114.1	−126.1	−89.80	−108	−103	−23.60
Water solubility at 25 °C (g/L)	0	1000	1000	1000	63.20	85	32.4	42.60
Anaerobic biodegradation probability (Biowin7)	–	0.89[a]	0.92[a]	0.94[a]	0.65[a]	0.67[a]	0.31[a]	0.32[a]
Research octane number (RON)	96	110	109.00	104	98	105	99.16[a]	98.56[a]
Flash point (°C)	−45	11.77	12.7	15.00	28.88	27.78	21.1	156.00

Vapor pressure at 25 °C (mmHg)	525	127.06	59.29	20.99	6.7	10.5	8.59	5.56
Viscosity at 25 °C (cP)	0.6	0.55	1.07	1.96	2.55	3.37	3.35 [a]	3.57[a]
Density at 25 °C (g/cm^3)	0.7	0.79	0.79	0.80	0.81	0.8	0.81	0.83
Kinematic viscosity at 25 °C (mm^2/s)	0.85	0.7	1.36	2.44	3.15	4.2	4.14[a]	4.3[a]
Surface tension at 25 °C (dyne/cm)	22	22.07	21.97	23.32	24.93	22.54	22.92	23.26
Heat of vaporization (kJ/kg)	–	1173.53[b]	918.60[b]	788.68[b]	708.31[b]	685.37[b]	535.36[b]	554.90[b]
Lower heating value (MJ/kg)	42.9	21.26	28	31.57	33.09	33.11	36.42 [a]	36.42 [a]
YSI values	40–60[c]	−36.9[c]	−31.1[c]	−22[c]	−13[c]	−6.5[c]	9.4[c]	11.4[c]

[a]Predicted via group contribution methods.
[b]298.15 K, Chickos and Acree, 2003. Enthalpies of Vaporization of Organic and Organometallic Compounds.
[c]Yield Sooting Index Database Volume 1, https://doi.org/10.7910/DVN/9FTNCK. Harvard Dataverse, V2.

Compression Ignition Engines

History and Use

As was the case for SI engines, multiple engineers and scientists are credited with early work on CI engines. However, the modern diesel engine can be exclusively traced to Rudolph Diesel's pioneering work in the mid-1890s. In 1893 he published the treatise *Theorie und Konstruktion eines rationellen Wärmemotors zum Ersatz der Dampfmaschine und der heute bekannten Verbrennungsmotoren*, later translated as *Theory and Construction of a Rational Motor*. From 1893 to 1897, he was issued a series of patents in Germany, Switzerland, the United Kingdom, and the United States on CI engine concepts, and in 1897 for a working prototype. Because of its high expansion ratio, lower pumping loss, lower ignition loss, and inherent lean burn operation, the diesel engine has the highest thermal efficiency of any practical internal or external combustion engine.

CI engines generate more torque than their SI counterparts because of higher compression ratio, longer piston power stroke, faster combustion, and the higher energy density of diesel fuel. To handle the higher torque and cylinder pressure, they must be built more rugged, hence their greater reliability and durability. All of these factors—greater torque, better fuel efficiency, longer life—make them best suited for medium and heavy-duty trucks, where the benefits outweigh the added cost. Most of the world's almost 400 million vehicles classified as trucks and buses are diesel powered. Diesels have other applications in off-road working vehicles (tractors, earth movers, bulldozers), locomotives, boats, ships, stationary power sources, and pipeline transport power supplies. They are pervasive and a critical component of the world's economy. Some current applications could be replaced by electric power, but most analyses show that this would be a bigger challenge than for SI engines. In recent decades, diesels have also found a large application in passenger cars and light trucks—40% of new car sales in Europe in 2015. However, because of emission concerns and emission cheating scandals, sales have dropped significantly, and the future is uncertain.

Operation and Fuel Needs

CI engines are classified by speed: high (>1000 rpm), medium (300–1000 rpm), and low (<300 rpm). Since all on-road and most off-road CI engines are high speed, this discussion will be about them. Fig. 9.2 is a

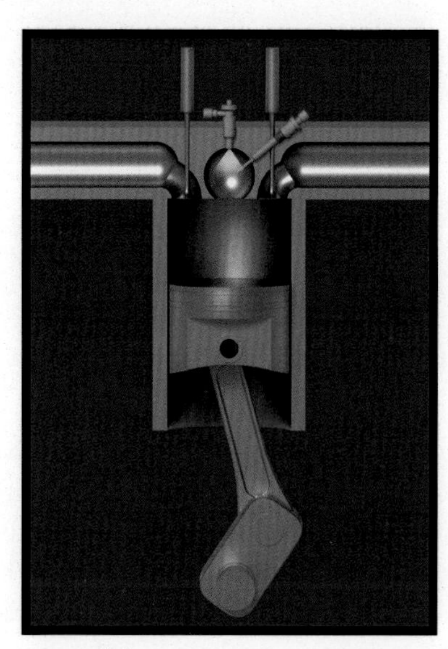

Fig. 9.2 Combustion process of the compression ignition engine.

schematic of the CI combustion process. The air is compressed by the upward movement of the piston to above ignition temperature and pressure for the fuel. The fuel is ignited instantly as it is injected directly into the cylinder. There is no risk of autoignition, so the engine can operate at higher compression ratio for higher torque and efficiency. The CI engine also has lower pumping loss, as it does not require a throttling valve, so efficiency is 20–40% better than for an SI engine. From an emissions perspective, an undesirable aspect of CI combustion is the non-premixed fuel-rich combustion, which results in a tradeoff between NO_x and soot emissions. Soot particles are formed in the fuel-rich combustion zones, and as they are burnt off at high temperature in near-stoichiometric reaction zones, NO_x is formed.

Important characteristics for diesel fuel are cetane rating, physical properties that affect spray characteristics, volatility, and emission properties. Because under high pressure injection conditions vaporization is limited by mixing and entrainment in the fuel spray rather than atomization, low-temperature ignition chemistry is the major factor determining cetane rating.

Diesel fuels have a higher boiling range than gasoline and jet fuel and are comprised of longer chain hydrocarbons, mostly branched and straight chain alkanes. The sooting propensity of alkanes, a major chemical component of diesel fuels, increases with carbon length, so traditional diesel fuel with typical carbon lengths of 12–20 has a high propensity to soot.

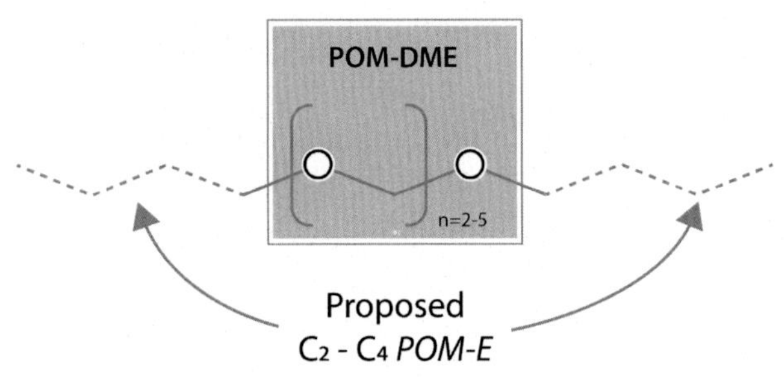

Fig. 9.3 Chemical structure of select polyoxymethylene ethers.

Oxygenated biofuel species when blended with a petro-diesel have the ability to reduce in-cylinder soot formation by altering kinetic pathways, diluting soot-forming hydrocarbons, and enhancing fuel-air mixing. Furthermore, many oxygenated species derived from bio-feedstocks offer kinetic pathways that can improve the ignition characteristics important for optimal CI operation relative to their petroleum counterparts. A good example of a biofuel oxygenate with these properties is polyoxymethylene dimethyl ether (POM-DME, Fig. 9.3, structure in shaded region). This class of molecules has the general structure CH_3-O-$(CH_2O)_n$-CH_3, where n is the number of oxymethylene units. Among POM-DME oligomers, those with $n = 2–5$ have garnered interest as diesel blendstocks, stemming from their favorable properties compared to conventional diesel, including the ability to enhance cetane numbers while offering significant soot reduction potential (50% reduction at 20% blending). Although POM-DMEs perform well in these regards, they suffer from lower heating values (LHVs), poor oxidative stability, and water solubility, all of which are issues arising from the molecules' high oxygen content and limit their suitability as a CI fuel blendstock. Polyoxymethylene ethers (POM-Es), in which the terminating functional groups are increased to C_2 to C_4, illustrated in Fig. 9.3 by the addition of the dotted structures, offer a way to address the water solubility, LHV, and oxidative stability concerns that plague POM-DMEs, while still maintaining a high potential for soot reduction and favorable ignition characteristics. Furthermore, POM-Es can be produced from alcohols derived from lignocellulosic biomass; these advanced POM-E biofuels demonstrate >50% greenhouse gas emission reduction compared to conventional diesel. Table 9.2 lists the properties of some attractive POM-Es compared to conventional diesel. They have dramatically lower sooting propensities and higher cetane numbers but lower heating values than conventional diesel.

Table 9.2 Properties of Three Polyoxymethylene Ethers Compared to Conventional Diesel.

Property	Conventional diesel	POM-DME (n = 3)	POM-DME (n = 4)	POM-DEE (n = 2)	POM-DEE (n = 3)
Melting point (°C)	0 to −20 (CFPP)	−43	−10	−45	−24
Boiling point (°C)	170–390	156	201	140	185
Density (20 °C, kg/L)	0.83	1.03	1.07	0.91	0.97
Cetane number	~55	124	148	64	80
LHV (MJ/kg)	~43	19.6	19	25.7	23.7
YSI	215	10.9	12	18.5	20.3

Jet Engines

History and Use

The first patent for a jet powered aircraft was issued by England to Frank Whittle in 1932, but compressor technology at the time was insufficient to allow fabrication of a working prototype. Hans Van Ohain started work on a similar concept in 1935 but with a practical approach that allowed him to overcome the compressor technology limitations of the time and develop and demonstrate a working prototype of the world's first jet plane in 1939. He was granted a US patent that year. Modern commercial jet planes essentially originate from this patent; almost all of them use an airbreathing design.

Commercial air travel demand is somewhat dependent on the world's economy, but on a normalized basis it has been growing at an annual rate of 5.8% since 1970 and is expected to grow at a 4.7% rate for the next 20 years. Although there might be some opportunities for electrification in short distance flights, a tremendous advance in the power to weight ratio of batteries would need to be made for long haul. Therefore, demand for jet fuels is likely to continue increasing for decades.

Operation and Fuel Needs

An airbreathing jet discharges fast moving gases to provide propulsion. Fig. 9.4 is a simplified schematic of a turbojet. Air is compressed by both a converging inlet and a compressor, fuel is added, the mixture is combusted, and the exhaust gases are passed through a turbine to run the compressor and through a nozzle to provide thrust. Most commercial jets have a turbofan in

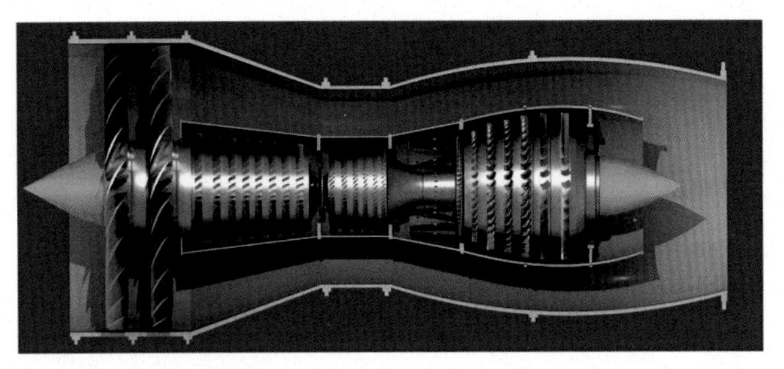

Fig. 9.4 Turbojet with a turbofan.

the inlet to bypass a portion of the air around the engine. This concept is more fuel efficient than a turbojet at subsonic speeds, where all commercial planes operate.

Conventional jet fuel is kerosene based, so it is defined by the kerosene boiling point in a distillation column, which falls between gasoline and diesel. Jet fuel has between 8 and 16 carbon atoms per fuel molecule. Unlike octane for gasoline or cetane for diesel, there is not a defined autoignition metric for jet fuel. Its important properties are flash point, freezing point, autoignition temperature, density, specific energy, and energy density. Given the safety implications, specifications on jet fuel are very stringent and tightly controlled.

Unlike SI and CI engines, which are reciprocating, a jet engine is a continuous flow device, which presents different challenges in flame stabilization to prevent blowout or flameout. Flame stabilization techniques vary by engine type but typically have a bluff body to impart a recirculation zone of hot combustion gases around the fuel injectors and compressed air inlet to prevent flameout. The fuel is not premixed with the combustion air, so similar tradeoffs of NO_x versus soot emissions occur as in CI. Flame conditions are quite different in a continuous flow jet engine, so engineering strategies to reduce emissions are different.

Oxygenated molecules are not allowed in jet fuel because of concern about degradation of cold temperature properties, which can occur at high altitude, along with decreased volumetric and mass energy densities. These problems translate into reduced range and load capacity, which is a significant detriment for commercial jet travel. Hence alternative jet fuels are required to meet the current specifications for kerosene-based fuel, which prohibits oxygenated fuel molecules. Many research groups have been working on fuels from renewable biomass feedstocks (biojet) that do not contain oxygen and meet jet fuel standards. Considerable progress has been made. Jet fuel blended with up to 50% biojet from a Fischer-Tropsch process was certified in August 2009. Biojet from hydroprocessing technologies such as hydro-treated esters and fatty acids (HEFA) and hydro-processed renewable jet (HRJ) were also studied extensively. Conversion of alcohol to jet fuel, called alcohol-to-jet (ATJ), was developed at commercial scale and flight-tested by the US Air Force in July 2012. Fuel produced by two recently proposed sugar-to-jet (STJ) processes, fermentation of sugars to hydrocarbons and catalytic conversion of sugars to fuels, have been developed in joint ventures by biofuel and oil companies. Hence progress on bio-based jet fuels has been very good, with many approaches achieving certification.[5]

An additional driver is the interest in reducing soot emissions from jet planes. Jet fuel is predominately *n*-alkanes, cycloalkanes, isoalkanes, and aromatic molecules. Aromatics are specified to be at a maximum 25% of volume because they have lower energy content and could lower combustion performance. Equally important, they have a higher propensity to generate PM emissions than the other compounds of jet fuel. Several organizations have expressed interest in lowering aromatic content in jet fuel and thus PM emissions. But this is complicated by current aircraft design, which requires a certain percentage of aromatics for sealant swell to maintain leak tightness of fuel tanks and lines. Hence a multi-tiered problem sets up where any jet fuel developed with lower aromatic content to reduce PM emissions would also have to address the sealant swell issue.

Analysis Needs

Extensive fuel analysis requires a fully equipped fuel characterization laboratory. To understand and quantify combustion properties, a fuel must be characterized at the molecular level. This understanding allows prediction of the performance and emission profile in developing advantaged biofuels for SI, CI, and jet engines. An important step for SI and CI fuels is determining the combustion characteristics, which involves determining the fuel ignition kinetics. Multiple approaches are available to measure ignition quality and delay. Here are some commonly used ones:

- Shock tubes are commonly used for investigating ignition delays for fuels. They are low cost but in general limited to measuring delay times of milliseconds.

- An ignition quality tester is a constant volume combustion vessel. It uses a set series of injections and pre-injections to reach steady state temperatures and determine the ignition delay.
- A research grade Advanced Fuel Ignition Delay Analyzer (AFIDA) is also a constant volume combustion chamber for mapping ignition delays, but allows for expanded experimental conditions and provides the capabilities to determine spray physics effects.

Another critical aspect of characterization is determining the emissions profile. Key components of an analysis system:

- Continuous measurement of gaseous emissions of total hydrocarbons, non-methane hydrocarbons, NO_x, carbon monoxide, carbon dioxide, and ammonia.
- Ability to measure both total PM and particle size. Particles below 10 μm in diameter pose the biggest health concern, as they penetrate deep into the lungs, decrease lung function, and enter the blood stream.
- Systems to measure unregulated emissions such as carbonyls, aldehydes, and ketones. These measurements require a gas chromatograph to allow speciation and quantification of C_1–C_{12} hydrocarbon emissions, including 1,3-butadiene and benzene, which are carcinogens under evaluation by some states for possible future regulation. A high pressure liquid chromatograph quantifies aldehydes, ketones, and carbonyls, known pulmonary irritants. A preconcentrator allows measurement down to parts per billion by volume; minute levels can cause pulmonary irritation.

If it is desirable to perform fuel characterization at the engine or vehicle level, then additional capabilities are required. Engine dynamometer test cells are set up for either light duty SI or heavy duty CI engines. Vehicle dynamometers simulate on-road driving under controlled conditions. Depending of the setup, the vehicle can be controlled by a computer or a driver. Driver control simulates as accurately as possible real-world conditions. Hence it is an important final step in quantifying a fuel's impact.

Close-up: Potential for Advantaged Biofuels in High-Efficiency, Low-Emitting Engines

There is a significant opportunity to improve spark-ignition engine efficiency through the combined application of more efficient engine design and operating strategies and better fuels.[6] The primary strategies are increased compression ratio, combined engine downsizing and turbocharging, operating at lower speed, cylinder deactivation, and direct injection of the fuel. Increasing compression ratio increases thermodynamic efficiency,

Close-up: Potential for Advantaged Biofuels in High-Efficiency, Low-Emitting Engines—cont'd

which increases the temperature and pressure of the unburned fuel-air mixture (end-gas), resulting in engine knock at high load. Thus, increasing compression ratio improves part-load efficiency but cannot improve high-load efficiency because it is limited by the knock resistance of the fuel. Engines exhibit lower efficiency when operated at light loads because of parasitic losses such as air pumping that result from the need to control the air-fuel ratio to be stoichiometric. Downsizing, turbocharging, and direct injection allow operation at higher load over a larger portion of the engine map. Operating at lower speed requires higher load to achieve the same power. Downsizing and down-speeding reduce friction, which improves efficiency. Cylinder deactivation at light load increases load and hence efficiency in the remaining cylinders. All strategies to operate the engine at higher load also cause higher temperature and pressure of the end-gas, and thus are limited by engine knock.

Spark-timing retard at high load mitigates knock at the expense of reduced efficiency due to less than ideal combustion timing. At very high load, fuel enrichment—operating at below stoichiometric air-fuel ratio—is applied to reduce knock via evaporative cooling and to control exhaust temperature to prevent engine and catalyst damage, but at the expense of high carbon monoxide and hydrocarbon emissions.

These strategies can all be pursued more aggressively, with less use of spark timing retard and fuel enrichment, if the engine is using a more knock resistant fuel. Properties that affect knock resistance include research octane number (RON), motor octane number (MON), octane sensitivity (OS = RON − MON), and heat of vaporization (HOV). Modern downsized boosted engines operate under conditions where increased RON and increased OS (or decreased MON) cause increased knock resistance. This effect has been described in terms of octane index (OI), a fuel's actual resistance to autoignition, where $OI = RON - K*OS$.[7] For downsized boosted engines, K is negative under most operating conditions, such that increasing S (or reducing MON at constant RON) increases OI.[8]

Regular gasoline in the United States today (a 10 vol% ethanol blend) has a RON of roughly 91 and OS of 8. Studies have identified many potential biofuel molecules with high RON.[9] However, only a subset of these blend non-linearly with petroleum refinery blendstocks as shown in Fig. 9.5 such that blends of 10–40 vol% have significantly higher RON than would be predicted by a volumetric or molar blending model. Blendstocks that demonstrate this synergism have the highest potential for enabling the introduction of higher efficiency spark ignition engines.[10] These include methanol, ethanol, isobutanol, prenol, diisobutylene, and alkyl furans, all of which also blend to increase OS.

Continued

Close-up: Potential for Advantaged Biofuels in High-Efficiency, Low-Emitting Engines—cont'd

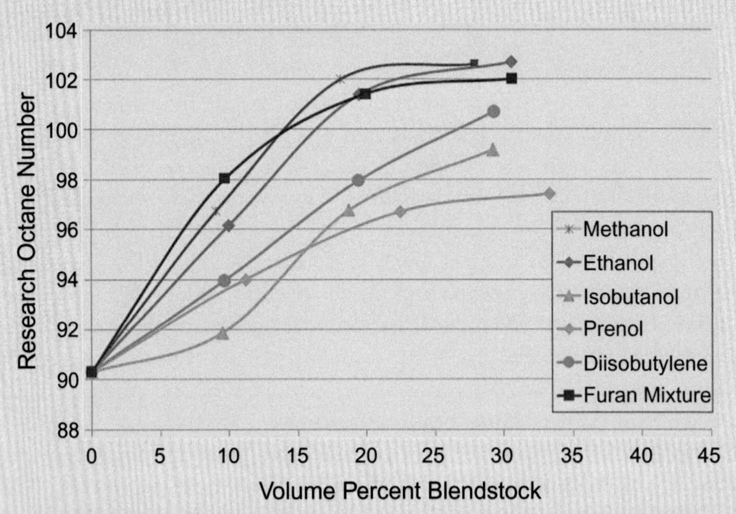

Fig. 9.5 Examples of blendstocks showing non-linear synergistic blending for research octane number.

While direct injection is necessary for high power density in downsized boosted engines, a second important effect is reduction in the fuel-air mixture temperature through evaporative cooling prior to intake valve closing.[11] Ethanol and other low molecular weight alcohols have a much higher HOV than hydrocarbon gasoline, so blending the two produces a significant increase in evaporative cooling and fuel knock resistance. Experiments and analyses have shown that charge cooling can range from roughly 15 °C for E0 to 30 °C at E50 and as high as 50 °C for E85 blends, and that these values are about 70% of the thermodynamic maximum possible cooling.[12]

Robert L. McCormick
National Renewable Energy Laboratory, Golden, CO, United States

References

1. Pope III CA, Ezzati M, Dockery DW. Fine-particulate air pollution and life expectancy in the United States. *N Engl J Med.* 2009;360(4):376–386.
2. Cohen AJ, Brauer M, Burnett R, et al. Estimates and 25-year trends of the global burden of disease attributable to ambient air pollution: an analysis of data from the Global Burden of Diseases Study 2015. *Lancet.* 2017;389(10082):1907–1918.

3. EPA. *History of reducing air pollution from transportation in the United States,* 2018
4. McEnally CS, Pfefferle LD. Sooting tendencies of oxygenated hydrocarbons in laboratory-scale flames. *Environ Sci Technol.* 2011;45(6):2498–2503.
5. Wang W-C, Tao L. Bio-jet fuel conversion technologies. *Renew Sust Energy Rev.* 2016;53:801–822.
6. Leone TG, Anderson JE, Davis RS, et al. The effect of compression ratio, fuel octane rating, and ethanol content on spark-ignition engine efficiency. *Environ Sci Technol.* 2015;49(18):10778–10789.
7. Kalghatgi G. *Fuel/Engine Interactions.* Warrendale, PA: SAE International; 2013.
8. Mittal V, Heywood JB. The relevance of fuel RON and MON to knock onset in modern SI engines. In: *Powertrains, Fuels and Lubricants Meeting, October 6–9. Rosemont, IL*; 2008.
9. McCormick RL, Fioroni G, Fouts L, et al. *Selection Criteria and Screening of Potential Biomass-Derived Streams as Fuel Blendstocks for Advanced Spark-Ignition Engines.* SAE International; 2017.
10. Farrell JT, Holladay J, Wagner R. *Co-Optimization of Fuels & Engines: Fuel Blendstocks With the Potential to Optimize Future Gasoline Engine Performance; Identification of Five Chemical Families for Detailed Evaluation.* Golden, CO: National Renewable Energy Lab. (NREL); 2018. https://doi.org/10.2172/1434413 NREL/TP-5400-69009; DOE/GO-102018-4970 United States NREL English.
11. Ratcliff M, Burton J, Sindler P, Christensen E, et al. *Effects of Heat of Vaporization and Octane Sensitivity on Knock-Limited Spark Ignition Engine Performance.* SAE Technical Paper 2018-01-0218. https://doi.org/10.4271/2018-01-0218.
12. Kasseris E, Heywood J. *Charge Cooling Effects on Knock Limits in SI DI Engines Using Gasoline/Ethanol Blends: Part 1—Quantifying Charge Cooling.* SAE Technical Paper 2012-01-1275. https://doi.org/10.4271/2012-01-1275.

Alternative Jet Fuels

Introduction

The *Annual Energy Outlook 2018* from the US Energy Information Agency shows that jet fuel accounted for 10% of the energy used for transportation in 2016. While the outlook for 2050 projects overall consumption in the transportation sector decreasing by 5%, the percentage of energy used in the aviation industry is expected to increase by 60%. Increased consumption translates into increasing CO_2 emissions and increasing financial risk associated with jet fuel price volatility and uncertainty in the future.

Jet fuel, like gasoline and diesel, is a blend of hydrocarbons. The carbon number distribution is from C_4 to C_{12} for gasoline and C_8 to C_{21} for diesel. The distribution for jet fuels is a narrower cut of diesel, C_8 to $C_{16.}$ In a petroleum refinery, crude oil is first sent to an atmospheric distillation unit to recover the "straight-run" light gases, naphtha, kerosene, and diesel products. Straight run kerosene-type jet fuels have a boiling range between 160 °C and 275 °C. The main components of jet fuels are linear and branched alkanes (paraffins), cycloalkanes (naphthenes), and aromatics. The relative proportions of these hydrocarbon components are varied to match the desired bulk properties of the fuel, specifically energy content, combustion characteristics, density, and fluidity.

Energy content and density are often correlated, but less dense fuels tend to have higher energy density on a weight basis while more dense fuels have a higher volumetric energy content. The aromatics content of jet fuels, specifically the naphthalenes, affects smoke (soot) formation. Smoke formation is controlled by optimizing engine design and proper air/fuel mixing to make sure that these small carbon particles are completely consumed, thus avoiding luminous combustion that causes hot spots in turbines, erosion of turbine blades, and particulate emissions. Viscosity and freezing point are the bulk fuel properties that affect jet fuel fluidity.

Trace components in jet fuel can also be manipulated or managed by using additives to control other physical properties such as lubricity, stability

Analytical Methods for Biomass Characterization and Conversion
https://doi.org/10.1016/B978-0-12-815605-6.00010-X

(thermal and oxidative), corrosivity, cleanliness, and electrical conductivity. Trace amounts of oxygen, sulfur, and nitrogen compounds act as lubricants to prevent wear on moving metal surfaces. These trace heteroatom compounds occur naturally in straight-run jet fuels but can be removed if the jet fuel cut is hydrotreated. Poor lubricity fuels can be improved by adding these components back into the fuel at very low concentrations (~10 ppm). Oxidative stability is required for maintaining fuel properties during storage. Antioxidants are added to inhibit the formation of gums and particulates that can clog fuel lines and filters. Formation of these unwanted impurities accelerates at higher temperatures, especially since jet fuels are used for heat exchange in fuel systems and turbine engines. Trace amounts of organic acids, mercaptans, and alkali metals are controlled to minimize corrosion. Water content and particulate loading define cleanliness.

Jet fuels go by many names that are often used interchangeably. In the United States, the commercial aviation industry uses a kerosene-type fuel called Jet A. The military uses JP-8, a slight modification of Jet A with additives for better corrosion and static protection. JP-4 and JP-5 are also military fuels. Outside of the Unites States, international carriers use Jet A-1, which has a lower freezing point than Jet A, making it more suitable for high altitude intercontinental flight.

Jet Fuel Standards

Fuel certifications for commercial aviation are set and maintained by ASTM International (formerly the American Society for Testing and Materials) and the United Kingdom Ministry of Defense (MOD). Aviation fuel specifications are the responsibility of subcommittee J of committee D-2, Petroleum Products and Lubricants, in ASTM. The Standard Specification for Aviation Turbine Fuels (ASTM D1655) covers Jet A and Jet A-1. The MOD maintains Defense Standard 91-91 for Jet A-1 outside of the United States.

Standards are governed by ASTM D1655 with cross reference to more than 30 associated standards for specifications of individual fuel properties. Table 10.1 gives some of the standards associated with bulk and trace properties, along with the targets for each property. Typically more than one standard applies for individual fuel properties, but collectively the standards dictate how fuels will be sampled and tested to conform to the requirements for composition, volatility, fluidity, combustion, corrosion, thermal stability, contaminants, and additives.

Table 10.1 Standard Test Methods and Ranges of Specific Properties for Jet Fuel Specification Testing in Accordance with ASTM D1655.

Property	Standard test method	Value
Distillation	ASTM D86—Distillation of Petroleum Products and Liquid Fuels at Atmospheric Pressure ASTM D2887—Boiling Ranges Distribution of Petroleum Fractions by Gas Chromatography ASTM D2892—Test Method for Distillation of Crude Petroleum (15-Theoretical Plate Column) ASTM D7344—Test Method for Distillation of Petroleum Products and Liquid Fuels at Atmospheric Pressure (Mini Method) ASTM D7345—Test Method for Distillation of Petroleum Products and Liquid Fuels at Atmospheric Pressure (Micro Distillation Method)	10% distillation 205 °C maximum Final boiling point 300 °C maximum
Thermal stability	ASTM D1660—Method of Test for Thermal Stability of Aviation Turbine Fuels ASTM D3241—Thermal Oxidation Stability of Aviation Turbine Fuels (JFTOT Procedure)	25 mmHg maximum filter pressure drop; < Code 3 tube deposits maximum
Density	ASTM D1298—Test Method for Density, Relative Density, or API Gravity of Crude Petroleum and Liquid Petroleum Products by Hydrometer Method ASTM D4052—Test Method for Density, Relative Density, and API Gravity of Liquids by Digital Density Meter	At 15 °C 775–840 kg/m^3
Viscosity	ASTM D445—Test Method for Kinematic Viscosity of Transparent and Opaque Liquids (and Calculation of Dynamic Viscosity) ASTM D7945—Test Method for Determination of Dynamic Viscosity and Derived Kinematic Viscosity of Liquids by Constant Pressure Viscometer	At −20 °C 8.0 mm^2/s maximum
Vapor pressure	ASTM D323—Test Method for Vapor Pressure of Petroleum Products (Reid Method) ASTM D5190—Test Method for Vapor Pressure of Petroleum Products (Automatic Method) ASTM D5191—Test Method for Vapor Pressure of Petroleum Products (Mini Method)	

Continued

Table 10.1 Standard Test Methods and Ranges of Specific Properties for Jet Fuel Specification Testing in Accordance with ASTM D1655—cont'd

Property	Standard test method	Value
Flash point	ASTM D56—Test Method for Flash Point by Tag Closed Cup Tester ASTM D93—Test Method for Flash Point by Pensky-Martens Closed Cup Tester ASTM D3828—Test Method for Flash Point by Small Scale Closed Cup Tester	38 °C minimum
Net heat of combustion	ASTM D240—Test Method for Heat of Combustion of Liquid Hydrocarbon Fuels by Bomb Calorimeter ASTM D1405—Test Method for Estimation of Net Heat of Combustion of Aviation Fuels ASTM D3338—Test Method for Estimation of Net Heat of Combustion of Aviation Fuels ASTM D4529—Test Method for Estimation of Net Heat of Combustion of Aviation Fuels ASTM D4809—Test Method for Heat of Combustion of Liquid Hydrocarbon Fuels by Bomb Calorimeter (Precision Method)	42.8 MJ/kg minimum
Freezing point	ASTN D2386—Test Method for Freezing Point of Aviation Fuels ASTM D5972—Test Method for Freezing Point of Aviation Fuels (Automatic Phase Transition Method) ASTM D7153—Test Method for Freezing Point of Aviation Fuels (Automatic Laser Method) ASTM D7154—Test Method for Freezing Point of Aviation Fuels (Automatic Fiber Optical Method)	Jet A 40 °C maximum Jet A-1 47 °C maximum

Naphthalenes content	ASTM D1840—Test Method for Naphthalene Hydrocarbons in Aviation Turbine Fuels by Ultraviolet Spectrophotometry	3 vol% maximum
Hydrogen content	ASTM D3343—Test Method for Estimation of Hydrogen Content of Aviation Fuels ASTM D3701—Test Method for Hydrogen Content of Aviation Turbine Fuels by Low Resolution Nuclear Magnetic Resonance Spectrometry	
Sulfur	ASTM D1266—Test Method for Sulfur in Petroleum Products (Lamp Method) ASTM D2622—Test Method for Sulfur in Petroleum Products by Wavelength Dispersive X-ray Fluorescence Spectrometry ASTM D3120—Test Method for Trace Quantities of Sulfur in Light Liquid Petroleum Hydrocarbons by Oxidative Microcoulometry ASTM D4294—Test Method for Sulfur in Petroleum and Petroleum Products by Energy Dispersive X-ray Fluorescence Spectrometry ASTM D4952—Test Method for Qualitative Analysis for Active Sulfur Species in Fuels and Solvents (Doctor Test) ASTM D5453—Test Method for Determination of Total Sulfur in Light Hydrocarbons, Spark Ignition Engine Fuel, Diesel Engine Fuel, and Engine Oil by Ultraviolet Fluorescence	0.3 wt% maximum
Mercaptan sulfur	ASTM D3227—Test Method for (Thiol Mercaptan) Sulfur in Gasoline, Kerosene, Aviation Turbine, and Distillate Fuels (Potentiometric Method)	300 ppmw max
Copper strip corrosion	ASTM D130—Test Method for Corrosiveness to Copper from Petroleum Products by Copper Strip Test	No. 1 for 2 h at 100 °C
Acidity	ASTM D3242—Test Method for Acidity in Aviation Turbine Fuel	0.10 mg KOH/g max
Existent gum	ASTM D381—Test Method for Gum Content in Fuels by Jet Evaporation	7 mg/100 mL maximum

Continued

Table 10.1 Standard Test Methods and Ranges of Specific Properties for Jet Fuel Specification Testing in Accordance with ASTM D1655—cont'd

Property	Standard test method	Value
Aromatics content	ASTM D1319—Test Method for Hydrocarbon Types in Liquid Petroleum Products by Fluorescent Indicator Adsorption ASTM D6379—Test Method for Determination of Aromatic Hydrocarbon Types in Aviation Fuels and Petroleum Distillates High Performance Liquid Chromatography Method with Refractive Index Detection	25 vol% max
Smoke point	ASTM D1322—Test Method for Smoke Point of Kerosene and Aviation Turbine Fuel	25 mm minimum
Electrical conductivity	ASTM D2624—Test Method for Electrical Conductivity of Aviation and Distillate Fuels	50–600 pS/m
Water separability	ASTM D3240—Test Method for Undissolved Water In Aviation Turbine Fuels ASTM D3948—Test Method for Determining Water Separation Characteristics of Aviation Turbine Fuels by Portable Separometer ASTM D7224—Test Method for Determining Water Separation Characteristics of Kerosine-Type Aviation Turbine Fuels Containing Additives by Portable Separometer	
Lubricity	ASTM D5001—Test Method for Measurement of Lubricity of Aviation Turbine Fuels by the Ball-on-Cylinder Lubricity Evaluator (BOCLE)	
Particulate matter	ASTM D2276—Test Method for Particulate Contaminant in Aviation Fuel by Line Sampling ASTM D5452—Test Method for Particulate Contamination in Aviation Fuels by Laboratory Filtration	1 mg/L maximum

Specifications for distillation address boiling range and carbon distributions that define the bulk fuel composition. Separate standards define the limits for individual components such as sulfur, naphthalenes, aromatics, and hydrogen. Density is another standard measurement for specifying bulk fuel composition. Volatility is related to standard measurements of vapor pressure and flash point, and fluidity is defined by specifications for viscosity and freeze point. Combustion properties are specified by the net heat of combustion standards that are used to measure energy content and smoke point, a measure of particulate formation during combustion. Naphthalene content is also correlated with particulate formation. Specific tests determine corrosion propensity, setting limits for degradation of metallic surfaces. Acidity and mercaptan content are also correlated with fuel system corrosion. The standard for thermal stability assesses the potential for deposit formation on heated surfaces and the formation of gums and solids during storage that can foul fuel distribution systems.

Specifications for contaminants set limits for the amounts of unwanted components that are delivered with the fuel and that form over time during distribution, handling, and storage. Aside from limitations on the various chemical components specified for composition, combustion properties, and corrosion, specifications for particulate matter (solids) and water content are used to determine contaminant loading. The solubility of water in jet fuel is small but not zero. It can end up as free water phase-separated on the bottom of a storage tank, in an oil-water emulsion, or dissolved in hydrocarbon liquids. Water content clearly affects energy content and ice formation at high altitude; but more importantly, free water provides a medium for microorganism growth that can lead to biofouling in fuel systems or produce chemical byproducts that lead to corrosion. Biological impurities produce higher than normal particulate loadings and promote soluble gum formation.

Additives are used to adjust specific properties such as lubricity, electrical conductivity, and freeze point. Sulfur additive packages have been developed for adjusting lubricity. Static dissipators are often added to adjust the electrical conductivity and make transport and storage safer. Emulsifiers can be added to remove water, and biocides to inhibit biological activity. These additives address specific fuel characteristics, but standards have been developed to ensure that there are no unintended consequences from individual additives or combinations of additives that affect other fuel specifications.

Alternative Jet Fuel Pathways

The commercial aviation sector faces increasing economic challenges from rising and volatile fuel costs and public and political scrutiny as environmental concerns about greenhouse gas emissions remain and the industry continues to grow. Jet fuel prices have increased to the point where they now account for nearly one-third of an airline's operating expenses. Fuel price volatility causes uncertainty in economic projections, and in the worst case can lead to service disruptions if airlines are faced with net operating losses.[1]

Even though the aviation industry is roughly 10% of the transportation sector, existing and projected growth means that increased fuel consumption will proportionally lead to increased CO_2 emissions. Commercial aviation accounts for an estimated 2–6% of global emissions.[2] In 2009, the International Air Transport Association (IATA) proposed to achieve carbon-neutral growth for the industry by improving fuel efficiency 1.5% annually between 2010 and 2020, maintaining carbon neutral growth beyond 2020, and reducing carbon emissions by 50% compared to 2005 levels by 2050.[3]

Aviation fuel efficiency has improved substantially, and only incremental improvements can be expected in the near future. In the longer term, translational technology development of new propulsion systems could dramatically improve efficiency, but it would take years and significant capital investment to replace the current fleet. Consequently, alternative fuels with low carbon footprint are the most feasible choice for improving the environmental performance of aviation transportation. However, these fuels are expected to be completely interchangeable and compatible with kerosene-type jet fuels, so no modifications to existing aircraft engines or fuel distribution systems are needed. This also requires that alternative fuels become cost-competitive and meet the stringent fuel specifications.

Several different pathways for producing biojet are under development at various stages of commercial viability.[2] These processes are grouped into four categories: oil-to-jet, gas-to-jet, alcohol-to-jet, and sugar-to-jet, listed in order of commercial readiness.[4]

The first generation and advanced biofuels conversion processes discussed in Chapter 3 can be tailored for production by each of these pathways. The front-end feedstock handling, pretreatment, and conversion unit operations are identical for the intermediates that can be upgraded into gasoline, diesel, or jet fuel. However, meeting the product specifications for jet fuel often requires customized downstream process configurations for

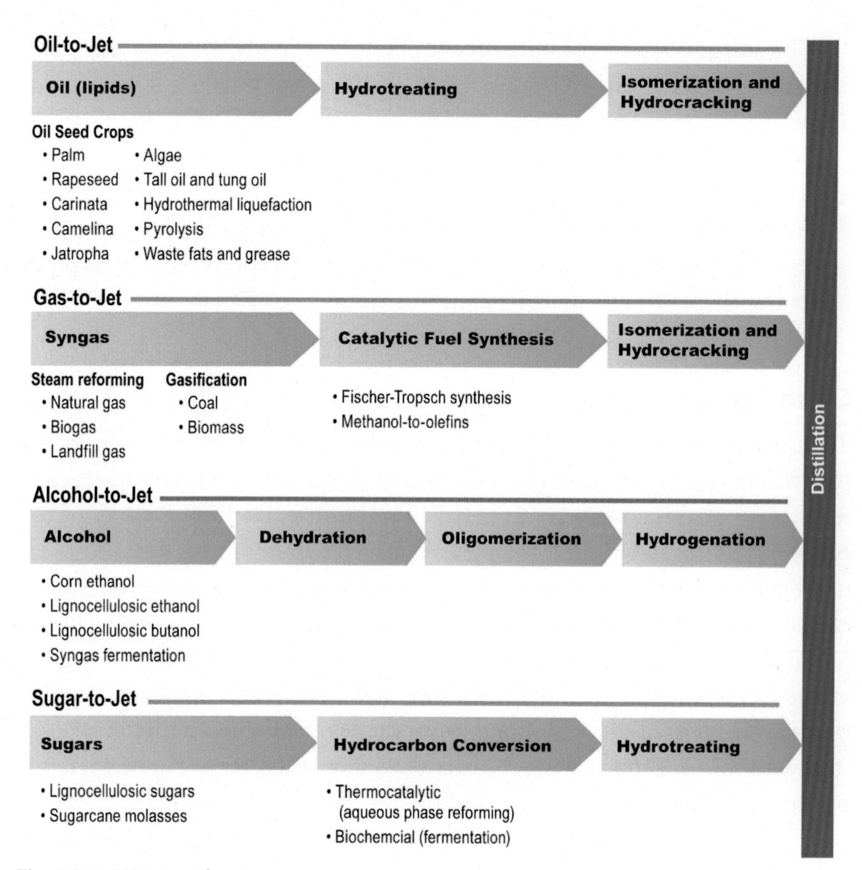

Fig. 10.1 Biojet pathways.

upgrading the variety of intermediates to product. Fig. 10.1 summarizes the process requirements for the four pathways. The following sections provide additional details.

Composition of alternative fuels from these pathways is a strong function of the feedstock and the production process. Since triglycerides in oil-to-jet are composed of three straight-chain esters, the resulting hydrotreated product is predominantly paraffinic with low aromatic content. Similarly, Fischer-Tropsch synthesis produces a high paraffin, low aromatic product. On the other hand, bio-oils made by biomass pyrolysis and hydrothermal liquefaction have high aromatic content, and hydrotreating these intermediates yields a product that contains aromatics and naphthenes (cycloparaffins). Consequently, biojet is often called synthetic paraffinic kerosene (SPK).

The sulfur content of alternative jet fuels is very low because the sulfur content of most biomass feedstocks is low and sulfur is usually removed during processing. For example, hydroprocessing in the oil-to-jet pathway removes heteroatoms (oxygen, nitrogen, and sulfur) by hydrodeoxygenation, hydrodenitrification, and hydrodesulfurization. Sulfur is a poison for Fischer-Tropsch catalysts, so syngas cleanup in integrated gasification processes always includes one or several sulfur removal steps to reduce gas phase sulfur impurities to less than 50 ppb.

Oil-to-Jet

This pathway involves technology developed and commercialized for recovering edible oils from oilseed crops in conjunction with hydrotreating technology that is a mainstay in petroleum refining. Both oil recovery from oilseeds and vegetable oil hydrotreating have been demonstrated and practiced commercially. This integrated process has reached commercial maturity and has been responsible for most of the alternative jet fuel produced and tested to date. Conventional biodiesel made by transesterification of triglycerides yields straight chain hydrocarbons and a glycerin byproduct. Hydrotreating converts all the components in the vegetable oils to fuel in a single step without forming any byproducts, thus increasing the carbon efficiency of the overall process.

The carbon chain length of the esters in the triglycerides, the degree of unsaturation, and the free fatty acid content are a function of the oilseed crop. The first step is to hydrogenate the triglycerides. Hydrotreating cleaves the fatty acid esters from the glycerol backbone and saturates and hydrodeoxygenates the separated fatty acid esters. The products from this step are propane and long-chain paraffins. Hydrocracking and hydroisomerization are performed to adjust the carbon distribution of the resulting hydrotreated jet fuel. The basis of the oil-to-jet pathway is hydroprocessing, so the resulting fuel is called hydrotreated renewable jet (HRJ) or hydrotreated esters and fatty acids (HEFA).

Candidate crops are rapeseed (with a variant known as canola), palm oil, and castor, to name a few. Non-food oilseed crops are receiving significant attention, particularly *Brassica carinata* (a relative of rapeseed, known as Ethiopian mustard) and jatropha. Carinata has a high glucosinolate and erucic acid content. The close phenotypic relationship with rapeseed makes carinata a suitable replacement. The drought and pest resistance of jatropha makes it a suitable replacement for palm oil in tropical regions.

Other sources of triglycerides for HRJ or HEFA processing are byproducts from the pulp and paper industry (tall oil and tung oil), algae, and waste fats (animal tallows) and greases. Tall oil hydrotreating is used commercially in Europe to produce renewable diesel and jet fuel; however, supplies are limited and tied to the pulp and paper industry. Producing large quantities of algae oil is still under development and not economical. Waste fats and greases are widely dispersed, making collection a challenge, and the volumes are too limited for sustainable fuel production.

Bio-crudes made by biomass pyrolysis or hydrothermal liquefaction are also hydroprocessed to make biojet. Bio-crude processes are still precommercial and require additional development to maximize yield and quality. These integrated thermochemical conversion processes are being developed to produce what is called hydrotreated depolymerized cellulosic jet (HDCJ) from hydroprocessed bio-crude intermediates.

Gas-to-Jet

This pathway relies on reforming or gasification technologies to produce a clean, cost-competitive syngas intermediate that can be catalytically converted to hydrocarbons by Fischer-Tropsch synthesis (FTS). The first commercial implementation of FTS, developed in 1926, was for gasoline from coal-derived syngas in Germany during World War II. Large scale FTS began in South Africa in 1952 when Sasol (South Africa Synthetic Oil Liquid) started making transportation fuels from coal. Initially the goal was to increase the valuation of the large domestic coal reserves, but later it became a necessity as the response to anti-apartheid sanctions and embargoes. Recent commercial FTS ventures rely on low-cost natural gas as the syngas precursor in large-scale gas-to-liquids plants that make gasoline, diesel, and hydrocarbon waxes.

Several feedstocks are used for syngas. Natural gas is the most abundant and cost-effective for a clean product. Steam methane reforming or partial oxidation can be used. Biogas and landfill gas are renewable methane sources, but they are highly distributed and therefore higher cost. Coal and biomass are used in integrated gasification processes. Syngas made from these solid fuels is more expensive than natural gas conversion, because extensive purification is needed to meet the stringent cleanliness targets for FTS.

The hydrocarbon product from FTS is essentially normal and isoparaffins, while the operating conditions of FTS dictate the hydrocarbon distribution in the products. High temperature FTS produces shorter chain hydrocarbons that are predominantly liquid. Low temperature FTS produces longer chain hydrocarbons that are solid waxes at room temperature. The liquids from high temperature FTS may require less processing, but the hydrocarbon yields of low temperature FTS are greater. Converting the waxes into fuel requires subsequent hydrocracking and isomerization. Both hydrocracking and hydroisomerization are regularly used in petroleum refining.

The technical feasibility of FTS is not an issue; however, the economic viability of making clean syngas for FTS is a challenge. Economies of scale can be leveraged for natural gas reforming or integrated coal gasification plants. The cost per unit output for these mega-plants is lower at large scale, but the financial risks are substantially higher because of the capital expenditure required. Environmentally friendly or lower cost waste feedstocks are usually available in lower volumes, thus requiring smaller plants. Smaller scale systems are less attractive because economies of scale cannot be realized. Therefore, the primary technical challenge facing the gas-to-jet pathway is downsizing the production and conversion processes while still making a cost-competitive fuel.

Alcohol-to-Jet

The feedstock is ethanol or butanol. The source of alcohol has no effect on jet fuel production, because the starting materials are single-component chemicals and not a blend of different components. Therefore, first generation corn or sugarcane ethanol and second-generation lignocellulosic ethanol are candidate sources, with cost being the primary differentiator. Bio-butanol technology is also under development, in which lignocellulosic sugars are fermented in a modified process that yields a mixture of acetone, butanol, and ethanol.

One approach for converting alcohol to jet fuel requires three catalytic steps. The first is alcohol dehydration over solid acid catalysts. Dehydration of ethanol produces ethylene; dehydration of *N*-butanol produces butenes. Oligomerization of ethylene produces a mixture of α-olefins with a range of carbon numbers. The jet fuel range olefins (C_8–C_{16}) are then hydrogenated to produce normal paraffins for diesel and jet fuel. The short chain olefins (less than C_8) can be separated and recycled back to the oligomerization reactor to increase jet fuel production. A similar process is used for butanol conversion; however, the products are a mixture of hydrocarbons with carbon numbers that increase by four starting with C_8. The jet fuel range products are primarily C_{12} and C_{16} paraffins.

The current market for blending 10% ethanol in gasoline is saturated, so ethanol production capacity remains flat. The demand for flex fuel vehicles that can use up to 85% ethanol is quite small, so an increase in the flex fuel vehicle fleet is not likely to increase future ethanol consumption. The US Environmental Protection Agency recently granted a waiver to use 15% ethanol blends in model year 2001 and newer vehicles. This provides an opportunity to increase ethanol consumption, but adoption has been slow. Using ethanol as a feedstock for jet fuel opens up an additional market beyond the current use as a blendstock in motor gasoline. Corn ethanol capacity could expand to meet this additional demand as the commercial interest in alcohol-to-jet increases. Additionally, jet fuel production is another option to avoid competition for market share between first- and second-generation ethanol production as lignocellulosic ethanol becomes a commercial reality.

The commercial readiness of the alcohol-to-jet technology is quite high. The production capacity for corn ethanol is nearly 15 billion gallons per year, and the individual ethanol conversion steps have been demonstrated at commercial scale in the petrochemical industry. The remaining developmental challenge is to combine these individual unit operations and demonstrate that the integrated process can produce cost-competitive jet fuel at commercial scale. Techno-economic analyses of first- and second-generation bioethanol conversion to jet fuel indicate that more technical improvements are needed to increase process efficiency and yield while reducing capital cost through process intensification.[5] One potential technical challenge that requires further evaluation is the impact of alcohol purity on the overall process. A major cost component in fuel-grade ethanol production is distillation and concentration to minimize the water content of the product. Energy cost and capital cost can be reduced if ethanol purity requirements are relaxed and ethanol/water blends can be made into fuel without additional process efficiency losses.

Sugars-to-Jet

This pathway consists of two technologies that convert first-generation (sugarcane molasses) and second-generation (lignocellulosic) feedstocks. One involves fermentation of sugars directly into C_{15} isoprenoids that can be hydrogenated to produce fuel. The other converts lignocellulosic sugars using aqueous phase reforming to make hydrocarbons that can be hydroprocessed to produce fuel. The sugar fermentation technology has been commercially demonstrated for making bioproducts but has not been demonstrated for large scale jet fuel production. Aqueous phase reforming has been demonstrated at pilot scale and requires additional development for scale-up.

Alternative Jet Fuel Certification

The increased interest in jet fuels from non-petroleum sources highlighted the need for a qualification and certification process to ensure that they meet the rigorous specifications established for conventional fuels as outlined in ASTM D1655. In 2009, a new standard for jet fuels containing hydrocarbons from Fischer-Tropsch synthesis was developed by ASTM to define the specifications for the properties and composition of blends containing synthetic hydrocarbons and conventional petroleum jet fuel.[6] Since then, ASTM D7566 (Standard Specification for Aviation Turbine Fuel Containing Synthesized Hydrocarbons) has been updated to include hydrocarbons produced by the other alternative pathways discussed in the previous section. An alternative fuel or blend is considered a "drop-in" replacement if it meets the property and composition specifications for ASTM D7566. Once a fuel or blend is designated as drop-in, it is reclassified as a conventional fuel (i.e., ASTM D1655 certified) in the aviation fuel market and infrastructure.

Meeting ASTM D7566 requires an extensive and rigorous approval process with multiple steps of fuel testing, component compatibility testing, and engine testing. To help prospective producers navigate this process, ASTM D4054 (Standard Practice for Qualification and Approval of New Aviation Turbine Fuels and Fuels Additives) was developed. The ASTM D4054 procedure is iterative and involves three comprehensive steps requiring increasing volumes of fuel.[6]

The first step is the test program, consisting of four tiers. Tier 1 requires up to 10 gal to determine the bulk properties, composition, and performance outlined in ASTM D1655. Tier 2 determines fit for purpose, that is, compatibility with the existing aviation fuel infrastructure in terms of storage, handling, distribution, and use in collaboration with the original equipment manufacturers (OEMs). It requires up to 100 gal, depending of the extent of testing. Tier 3 consists of component and rig testing to evaluate fuel system performance, combustion performance, and corrosion properties. It can require from 250 to 10,000 gal. Tier 4 is long-term engine testing to evaluate steady state performance and the impact, if any, of trace components on engine wear and maintenance. It requires commercial quantities, hundreds of thousands of gallons.

The second step is the OEM internal review, where equipment manufacturers evaluate the test results.

The third step requires a vote of the ASTM membership after another extensive review of the test program data and OEM review by a diverse group of stakeholders. Fuels approved to date are listed in Table 10.2. Fischer-Tropsch fuels (FT-SPK), hydrotreated vegetable oils (HEFA-SPK), and alcohol-to-jet fuels (ATJ-SPK) are approved for commercial use in blends up to 50% with petroleum-derived jet fuel. The hydroprocessed isoprenoids made from fermentation of sugars (HFS-SIP) are approved for blends up to 10%. Many fuels from other pathways are still being considered for certification.

Table 10.2 Alternative Jet Fuels Approved for Use or Pending Approval.

Fuel	Certification year	Blend ratio	Feedstocks
Fischer-Tropsch Synthetic Paraffinic Kerosene (FT-SPK)	2009	50%	Biomass (including MSW), coal, and natural gas
Hydroprocessed Esters and Fatty Acids Synthetic Paraffinic Kerosene (HEFA-SPK)	2011	50%	Plant and animal fats, oils and grease
Hydroprocessed Fermented Sugars to Synthetic Isoparaffins (HFS-SIP)	2014	10%	Sugars
Fischer-Tropsch Synthetic Paraffinic Kerosene with Aromatics (FT-SPK/A)	2015	50%	Biomass (including MSW), coal, and natural gas
Alcohol-to-Jet Synthetic Paraffinic Kerosene (ATJ-SPK)	2016	50%	Starches, sugars, lignocellulosic biomass
Hydro-deoxygenation Synthetic Kerosene (HDO-SK)	Pending		Sugars and lignocellulosic biomass

Continued

Table 10.2 Alternative Jet Fuels Approved for Use or Pending Approval—cont'd

Fuel	Certification year	Blend ratio	Feedstocks
Catalytic Hydrothermolysis Synthetic Kerosene (CH-SK)	Pending		Renewable fats, oils, and grease
Hydro-deoxygenation Synthetic Aromatic Kerosene (HDO-SAK)	Pending		Sugars and lignocellulosic biomass
High Freeze Point Hydroprocessed Esters and Fatty Acids Synthetic Kerosene (HFP-HEFA-SK)	Pending		Renewable fats, oils, and grease
Alcohol-to-Jet Synthetic Kerosene with Aromatics (ATJ-SKA)	Pending		Sugars and lignocellulosic biomass
Hydroprocessed Esters and Fatty Acids Synthetic Paraffinic Kerosene (HEFA-SPK)	Pending		Algae oil
Integrated Hydropyrolysis and Hydroconversion (IH^2)	Pending		Biomass (including MSW)
Hydrotreated Depolymerized Cellulosic Jet (HDCJ)	Pending		Biomass

Commercial Scale Production

Technology advancement and fuel certification are two critical components for developing alternative jet fuels, but to be commercially viable, the entire value chain needs to be optimized. Feedstock resource availability and logistics are crucial. Low cost and sustainable feedstocks are key for making cost-competitive, environmentally friendly products at commercial scale. Fuel offtake agreements are also required to reduce economic risk and encourage investment in new technology. A stable policy and regulatory environment is also required to present stable market signals that attract investment.

There is a unique commercial opportunity because the aviation industry is committed to reducing greenhouse gas emissions and improving economic stability by minimizing fuel cost volatility. Both environmental and economic benefits create a market pull that has encouraged several public-private partnerships that catalyze industry growth.

The US government made two large initiatives to provide market opportunities. The Department of the Navy created a Biofuels Initiative in 2012 under the Defense Production Act Title III to assist the development and support of a sustainable commercial biofuels industry. This led to financial assistance for three commercial demonstrations with the potential to produce biofuels for military applications. The Department of Agriculture, in partnership with Airlines for America (an aviation trade organization) and Boeing, established the Farm to Fly Initiative in 2010 to support the development of a commercially viable domestic industry. The goal of Farm to Fly was to produce 1 billion gallons of sustainable biojet by 2018. A major emphasis of this initiative is developing non-food oilseed crops and agronomics to maximize greenhouse gas reduction potential at a scale that provides commercially relevant quantities of non-edible oils as feedstocks.

The Commercial Aviation Alternative Fuels Initiative (CAAFI; see the close-up below) was established in 2006 to enhance energy security and environmental sustainability for the industry through alternative jet fuel use. CAAFI is a coalition of commercial airlines, OEMs, alternative fuel producers, researchers and developers, and US government agencies. The goal is to cultivate relationships among these stakeholders, share data, identify resources, and recommend additional research, development, and deployment needs. Partnerships are being established as commercial airlines align themselves with alternative fuel developers, as evidenced by offtake agreements.

Since these early initiatives, several commercial processes have been developed.[4] Honeywell UOP was one of the first companies to offer commercial technology based on hydrotreating vegetable oils for diesel and jet fuel (HEFA-SPK). Neste Oil, a Finnish company, has developed and commercialized a process based on vegetable oils called NExBTL. Fischer-Tropsch and alcohol-to-jet methods are moving toward commercial production.

Increasing production has led to many successful commercial flights using alternative blends. IATA statistics show 100,000 flights since the first one in 2008.[3] The literature contains references to specific flights,[7] but the pace at which new flights are occurring is impossible to keep up with, suggesting a strong market pull for alternative jet fuels in the future.

Close-up: The Commercial Aviation Alternative Fuels Initiative (CAAFI)

The commercial aviation industry is acutely sensitive to the price volatility of jet fuel. A large fraction of any airline's operating cost is for fuel, so profit decreases as fuel cost increases. The financial health of many airlines was adversely affected when petroleum prices rapidly ran up and peaked in 2008, causing consolidation in the industry. Some airlines merged while others filed for bankruptcy. At the same time, greenhouse gas emissions from the transportation sector were increasingly being scrutinized. The economic and environmental concerns associated with the airline industry can be addressed by improving the fuel efficiency of air travel and seeking alternatives to petroleum-based jet fuel.

The threat of fuel supply disruption and price volatility, in conjunction with environmental concerns, led to the establishment of CAAFI in 2006 to promote the development of cost-competitive alternative jet fuels that have equivalent performance and safety as petroleum jet fuel but with lower greenhouse gas emissions. CAAFI members include airlines, aircraft and engine manufactures, energy producers, researchers, and US government agencies. This coalition of stakeholders is focused on reducing the carbon footprint of air travel as the number of travelers and miles traveled increases into the future. The International Air Transport Association has an aggressive goal to reduce net CO_2 emissions by 50% relative to 2005 levels by 2050. The US Federal Aviation Administration was committed to use 1 billion gallons of renewable jet fuel by 2018.

CAAFI does not support specific alternative fuel technologies or promote specific feedstocks. Rather, they focus on the commercial deployment of sustainable alternative fuels. They provide information, connect stakeholders across the highly integrated supply chain with renewable fuel producers and developers, and assist policies and government initiatives that support the development of alternatives.

CAAFI is organized in four areas: research and development, fuel certification and qualification, sustainability, and business.

The Research and Development Team stays up to date with current activities on developing new pathways for alternative fuel production.

Once a pathway reaches the point where commercial viability seems possible, the Fuel Certification and Qualification Team follows the industry evaluation process for ensuring that the fuel meets the performance criteria and safety requirements for commercial aviation. Turbine aircraft engines require fuel with high energy content to meet high-performance demands. Jet fuel also requires good flow characteristics and thermal stability to achieve high performance over a wide range of demanding operating conditions from the ground to high altitude. By working with ASTM

Close-up: The Commercial Aviation Alternative Fuels Initiative (CAAFI)—cont'd

International (formerly the American Society for Testing and Materials), alternative fuel developers gain certification that their products are safe alternatives to petroleum-based fuel and can be used commercially without any new or modified equipment or infrastructure. ASTM D4054 provides guidance and requirements for the technical evaluation of alternative fuels. If a fuel meets the ASTM D4054 standard, it can be added to the D7566 specification as an approved drop-in replacement.

The Sustainability Team follows the development of lifecycle assessments (LCAs) for approved alternative fuel pathways to determine if they meet the greenhouse gas emission reduction potential for environmental sustainability. The social sustainability (for example, land use change and rural development) and economic risks associated with commercial deployment are also assessed.

This supports the Business Team in evaluating business models and identifying deployment opportunities to ensure commercial success.

References

1. Gegg P, Budd L, Ison S. The market development of aviation biofuel: drivers and constraints. *J Air Transp Manag*. 2014;39:34–40.
2. Wang W-C, Tao L. Bio-jet fuel conversion technologies. *Renew Sust Energ Rev*. 2016;53:801–822.
3. de Juniac A. *IATA Annual Review 2018*. International Air Transport Association; 2018.
4. Mawhood R, Gazis E, de Jong S, Hoefnagels R, Slade R. Production pathways for renewable jet fuel: a review of commercialization status and future prospects. *Biofuels Bioprod Biorefin*. 2016;10(4):462–484.
5. Tao L, Markham JN, Haq Z, Biddy MJ. Techno-economic analysis for upgrading the biomass-derived ethanol-to-jet blendstocks. *Green Chem*. 2017;19(4):1082–1101.
6. Zhang C, Hui X, Lin Y, Sung C-J. Recent development in studies of alternative jet fuel combustion: progress, challenges, and opportunities. *Renew Sust Energ Rev*. 2016;54:120–138.
7. Gutierrez-Antonio C, Gomez-Castro FI, de Lira-Flores JA, Hernandez S. A review on the production processes of renewable jet fuel. *Renew Sust Energ Rev*. 2017;79:709–729.

Additional Reading

Chuck C. *Biofuels for Aviation: Feedstocks, Technology and Implementation*. Waltham, MA: Academic Press; 2016.

CHAPTER ELEVEN

Biomass-Based Products and Chemicals

Introduction

The concept of producing materials and chemicals from biomass is nothing new. Food, animal feed, and fiber have been societal mainstays for generations. Food can be considered the original bioproduct, transforming atmospheric CO_2 and water into stored chemical energy through biological photosynthesis. The agricultural industry supports food crops for human consumption while some grains and agricultural residues are used for animal feed. Woody biomass can be considered the original biofuel, as it has been historically used for heat and light. The wood products industry currently transforms woody biomass into timber, pulp, and paper. In this case, the notion that biomass itself is the bioproduct has evolved to where biomass is the raw material for a process that converts it into higher value products in an integrated biorefinery.

Historically a pulp and paper mill is the first example of a biorefinery. There are several pulping processes (mechanical, sulfite, organosolv), but the kraft process produces the highest quality cellulose fibers. Softwood logs, typically southern yellow pine, are delivered to the mill site, where they are debarked and chipped. The woodchips are loaded into digesters with pulping chemicals (Na_2S and NaOH) to dissolve lignin and separate out the cellulose fibers. The fibers are dewatered, pressed, dried, and bleached for papermaking. The dissolved lignin and spent pulping chemicals make up a stream called black liquor. Initially, black liquor was disposed of, usually into a nearby river. As the scale of the kraft process increased and raw materials became more expensive, discharging black liquor became uneconomical as well as unacceptable. With the development of the recovery boiler, black liquor combustion provided process heat for the mill (bioenergy), and the pulping chemicals were recovered and recycled to the digestor. Hog fuel boilers were developed to provide additional process heat and consume

Analytical Methods for Biomass Characterization and Conversion
https://doi.org/10.1016/B978-0-12-815605-6.00011-1

the bark removed from the logs. In the sulfite pulping process, lignosulfonates are recovered from the spent pulping liquor and used in making concrete and cement.

Corn ethanol production is another example of a commercial biorefinery. Ethanol is produced from corn in either a dry or wet milling process. Several co-products can be recovered, the primary one being the stillage (corn fiber, gluten, and residual starch) remaining after the starch is separated from the kernel. This is dried for animal feed, called distillers dried grains. Others are corn oil and corn syrup. In the early days of the corn ethanol industry, these co-products were considered byproducts. Now robust markets have developed, and they generate significant revenue.

Maximizing the value of the input raw material is a basic strategy of the petroleum industry. Crude oil is principally a mixture of hydrocarbons with a broad range of boiling points and a wide molecular weight distribution. Trace components include sulfur and nitrogen compounds and heavy metals. A petroleum refinery is a series of separation and upgrading units that remove trace components and fractionate the input crude oil into streams that are uniquely processed to produce intermediates or products. Fig. 11.1 is a schematic of a simple refinery.[1] Gasoline, diesel, and jet fuel make up the largest volume of products, but the highest value per unit volume is generated by chemical products. Petrochemicals account for only 3% of the barrel of crude but create the same total value in the supply chain as the 70+% of the barrel used to make fuel. Complex and highly sophisticated linear programming models optimize the flow of intermediates and products to maximize revenue in response to changing market conditions.

The petrochemical industry uses primarily refinery gases and naphtha as feedstocks for making the building block molecules that are the foundation for a wide variety of chemical products. Steam cracking of refinery gases produces olefins, mainly ethylene and propylene, that can be used to make plastics (polyethylene and polypropylene) or other building block compounds (ethylene oxide and propylene oxide). The list of chemical products from these oxide intermediates expands into many other applications. Recovering butane from refinery gases is a separate stream that yields another suite of products (synthetic rubbers and polymers) and other chemical intermediates (butadiene, isobutylene, and *n*-butane). Recovering aromatics such as benzene, toluene, and xylenes from naphtha supports yet another section of the industry. Benzene is converted to building blocks such as maleic anhydride and cumene for making a wide range of polymers and plastics. Toluene is used for solvents and polyurethane production.

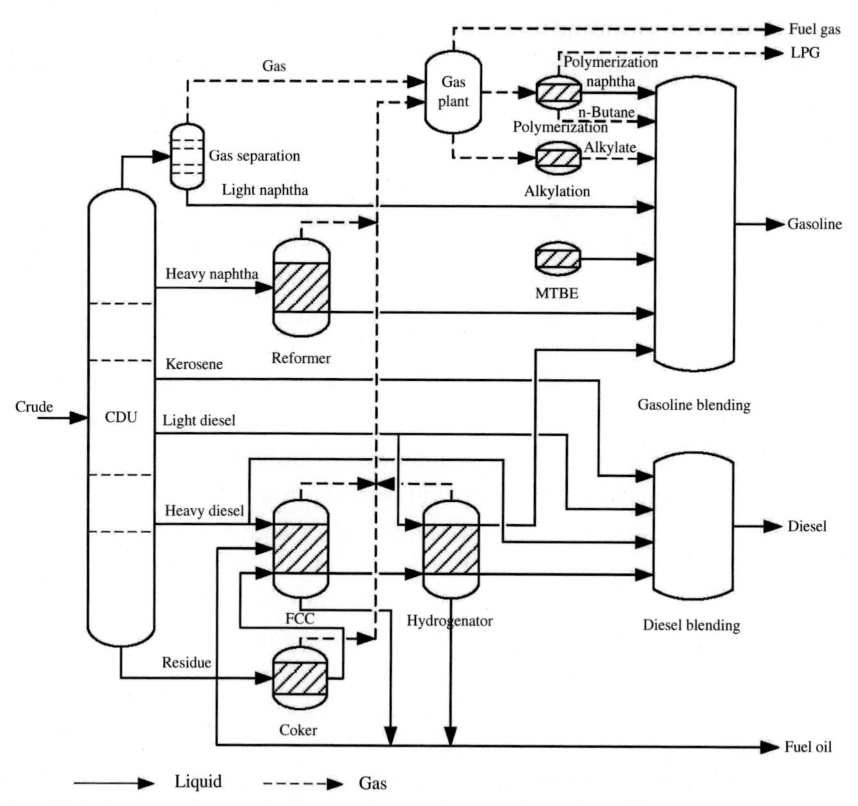

Fig. 11.1 Schematic of a simple petroleum refinery. *(Reproduced from Rong G, Zhang Y, Zhang J, Liao Z, Zhao H. A robust engineering strategy for scheduling optimization of refinery fuel gas system.* Ind Eng Chem Res. *2018;57:1547-1559.)*

Xylenes are precursors to several polymers, the most recognizable being polyethylene terephthalate, the plastic used for beverage containers.

Imagine treating biomass in a similar manner to petroleum. Developing highly efficient and cost-effective integrated processes could enable biomass conversion to both fuels and high-margin bioproducts. Biomass fed into a biorefinery would be deconstructed and separated into different process streams, and each stream could be independently upgraded to biofuels or intermediates to feed a bioproducts industry. The inherent functionalized nature of biomass offers a unique opportunity for the production of bio-based oxygen-containing intermediates and chemicals that are not easily synthesized from petroleum.[2, 3]

The physical properties of biomass pose unique challenges for developing biorefineries. Bulk density is much lower than that of petroleum, and as a solid feedstock, significant preparation and handling are needed. It must be dissolved and depolymerized, liquified, or gasified to produce process streams that can be separated and upgraded. The high oxygen content of biomass and its intermediates has much lower energy density than petroleum; consequently, the scale of refineries is almost two orders of magnitude smaller than for petroleum refineries. The output from a 2000 dry tonnes per day biorefinery is on the order of 5000 barrels; for petroleum refineries, it is 100,000–500,000. Limited economic benefits can be gained from economies of scale for biorefineries. The co-production of chemicals could enhance the viability of biomass-to-liquid fuel and afford a higher return on investment.[4] A study on the economic impact of co-production indicated that the cost of producing biofuels could be reduced by about 30%.[5] Regardless of the strategy, the integration of biofuels with high-margin co-products is an opportunity for additional revenue.

To recover as much value as possible, either higher value applications of biomass as a raw material must be sought or the biomass must be deconstructed into more basic components and those components valorized. Understanding the physiochemical structure of biomass and how to unpack the bulk components and recover the trace components provides insight into the chemical functionality that can be exploited for potential bioproducts and specialty biochemicals.

The bulk components of terrestrial biomass are cellulose, hemicellulose, and lignin. The trace components include non-structural organics collectively called extractives and inorganic constituents called ash. The extractive and ash contents of various types of biomass can be very different. While ash has little value, extractives such as turpentine, rosin, and other fatty acids, waxes, sterols, proteins, tannins, terpenes, and lipids have high value.

Natural Products

A well-known example is the exudate of *Hevea brasiliensis*, the rubber tree. Latex, or natural rubber, is an emulsion of water and polymers of isoprene that is collected and processed. Similarly, food oils are recovered from oilseed crops, as discussed in Chapter 3. Many natural products are also used for flavorings or fragrances. Vanilla beans contain a high concentration of vanillin, a common food additive. Essential oils from peppermint, spearmint, clove, and eucalyptus are used as flavorings and fragrances but have also been sought for medicinal and biocidal purposes. Discovering compounds that are biologically produced in nature for medicinal use is as old as civilization. Salicylic acid ingested by chewing the bark of the willow tree is one of the oldest analgesics; it led to the synthesis of acetylsalicylic acid—aspirin. Morphine comes from the opium poppy. A more recent example is the anticancer drug Taxol, a terpenoid extracted from the Pacific yew tree.

A primary feedstock in the wood products industry for timber, pulp, and paper is softwood, principally pine. Only about half of the green wood delivered to sawmills is converted to saleable timber,[6] and only the cellulose fraction is separated for pulp (~25 wt% on a wet basis, 50% on a dry basis). There is a significant opportunity to maximize value by looking for alternative uses of underutilized raw material or co-products in existing processes. An excellent example is the pine chemicals industry that extracts black liquor soap skimmings from the kraft pulping process, makes crude tall oil therefrom, refines that to isolate rosin and fatty acids, and finally produces high value specialty chemicals used in publication inks, adhesives, paints, soaps and detergents, metalworking fluids, paper chemicals, synthetic rubber, asphalt additives, and mining chemicals.

A great deal of literature is available on the isolation and derivatization of chemicals from pine, including turpentine, fatty acids, rosin acids, and lignin chemicals. These have been used in some form since preindustrial times, when they were sought for sealants in building construction and waterproofing of sailing vessels. Pine chemical use in sailing vessels was part of the foundation of the naval stores industry. Modern pine chemical processes focus on distillation of co-product liquors generated in the kraft pulping process (turpentine and tall oil), distillation of turpentine and crude rosin from tapping of standing pine trees (oleoresin), or solvent extraction of pine stumps. As a result, much is known about the composition of chemicals available from pine trees, purification methods, and conversion into industrial chemicals.[7–9]

Biomass Refineries

Cellulose is the most abundant biomass component, roughly half of the organic carbon in the biosphere. The cellulose polymer consists of repeating cellobiose units. Cellobiose is a disaccharide made up of two glucose molecules linked by a $\beta(1 \rightarrow 4)$ bond, with the formula $C_{12}H_{22}O_{11}$. Cellobiose hydrolysis produces two glucose molecules. Hemicellulose is the next most abundant, an amorphous material with a random heteropolymer containing C_5 (xylose and arabinose) and C_6 (mannose, glucose, and galactose) sugars. Lignin is the third most abundant, an aromatic three-dimensional polymer with phenylpropane building blocks known as monolignols.

Conversion technologies were covered in Chapter 3. The first step in any integrated biorefinery is to convert biomass into an intermediate that can be upgraded into biofuels or bioproducts. The lowest value streams can be burned for energy or disposed of as waste. The objective of pretreatment in a process is to separate lignin and deconstruct cellulose and hemicellulose to produce sugars. Pretreatment may take more than one step, such as fractionation and hydrolysis, to produce a variety of C_5 and C_6 sugars. Biocatalysts can be developed to convert specific sugars into targeted products other than ethanol with high efficiency and selectivity. Thermochemical processes convert all components into an intermediate in a single step. The intermediate in biomass gasification is syngas, and the intermediate for direct liquefaction technologies (pyrolysis and hydrothermal liquefaction) is bio-oil. Clean syngas can be catalytically converted into a variety of products. Recovering bioproducts and biochemicals from bio-oils depends on efficient separation technology to concentrate targeted compounds that are present in low concentration in a very complex mixture. Biochemical processes offer greater potential for efficiently recovering individual chemicals in high yield. Thermochemical processes are less sensitive to feedstock compositional variation, so intermediates can be produced with high efficiency, but the specificity of individual chemicals is very low. One could argue that biochemical processes are better suited for bioproducts and biochemicals whereas thermochemical processes are better suited for biofuels, but the concept of recovering bioproducts to improve economics and enable biofuels production is valid regardless of the technology.

Bioproducts and Biochemicals From Lignocellulosic Sugars and Lignin

The first step in an integrated lignocellulosic biorefinery is to deconstruct biomass into its three primary components: cellulose, hemicellulose, and lignin. Lignocellulosic sugars are then produced by hydrolyzing cellulose and hemicellulose. Previous chapters have discussed lignocellulosic ethanol from sugars. Ethanol dehydration also produces olefins that are precursors to jet fuel and other bio-based chemicals and polymers. Acetone-butanol-ethanol fermentation makes these three chemicals from sugars.

In additional to being a source for biofuels, lignocellulosic sugars are building blocks for higher value chemicals and bioproducts. A seminal report titled "Top Value Added Chemicals from Biomass" by Werpy and Petersen in 2004 details the results of a screening process to determine the potential for producing building block chemicals from these sugars.[10] They identified 12 candidate compounds with multiple functional groups that can be transformed into precursors for commercially viable products: 1,4-diacids (succinic acid, fumaric acid, and malic acid), 2,5-furan dicarboxylic acid, 3-hydroxy propionic acid, aspartic acid, glucaric acid, glutamic acid, itaconic acid, levulinic acid, 3-hydroxybutyrolactone, glycerol, sorbitol, and xylitol. Fig. 11.2 shows their structures.

Biological processes such as aerobic and anaerobic fermentation using biocatalysts (yeast, fungi, or enzymes) convert lignocellulosic sugars into

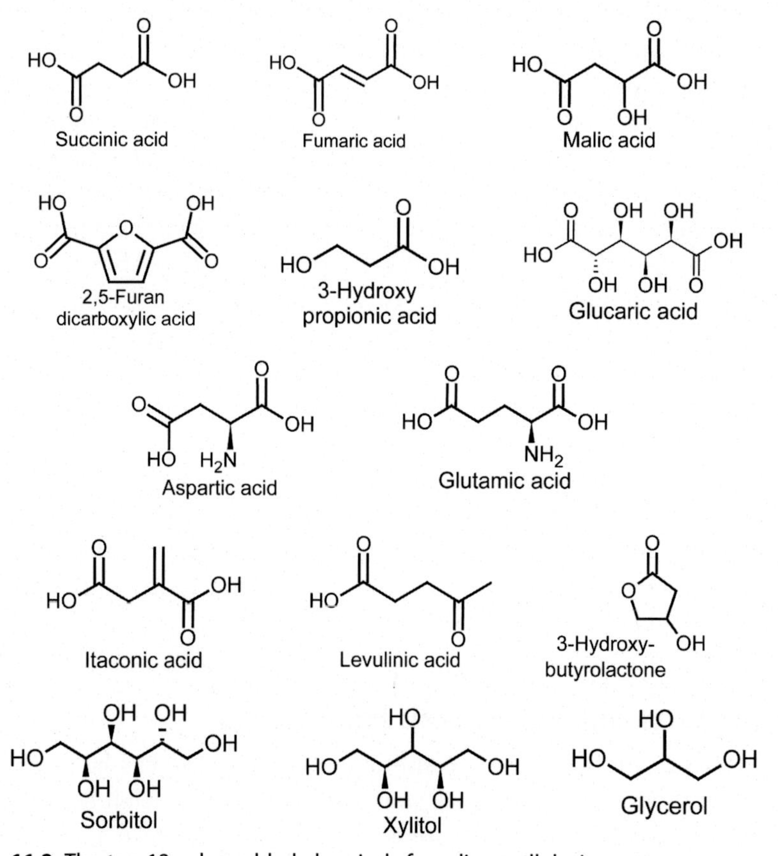

Fig. 11.2 The top 12 value-added chemicals from lignocellulosic sugars.

the building block molecules. Once produced, these molecules can be converted into derivatives or intermediates in various industrial chemical processes. For example, anaerobic fermentation of glucose produces succinic acid. Well-known reduction chemistries can be applied to succinic acid to produce tetrahydrofuran, 1,4-butanediol, or gamma butyrolactone. These intermediates are precursors for solvents and polymer fibers. Reductive amination of succinic acid yields pyrrolidinone derivatives used to make solvents and polymers. Isikgor and Becer have published a comprehensive review of the biochemicals (more than 200) that can be made from the platform compounds resulting from biomass-derived C_5 and C_6 sugars.[11]

The third most abundant biopolymer that can be separated from biomass is lignin. Traditionally it has been burned for heat and power generation in biorefineries. Higher value uses have been the focus of research and development for decades. The chemistry and structure of this alkyl-aromatic polymer varies considerably depending on the type of biomass it is isolated from. This variability makes lignin utilization a challenge.

There is an old saying in the biofuels industry that you can make anything from lignin except money but that is slowly changing. With little or no modification, it is a renewable carbon source for carbon fibers, polymer additives, adhesives, and resins. It behaves more like a plastic than the biomass starting material from which it was separated. Consequently, it can be melted and spun into fibers. Heat treating the aromatic backbone of lignin during the spinning process produces carbon fibers with a graphitic structure. Lignin is also a source of phenol in resins and adhesives and an additive in concrete and cement.

Recent advances in the understanding of lignin's chemical structure and chemistry have facilitated discovery of new thermocatalytic and biological processes that point to pathways for higher value products.[12] Depolymerization through selective catalytic hydroxylation or hydrogenolysis yields monomeric aromatics such as benzene, toluene, xylenes, phenols, and methoxyphenols. Selective bioconversion of aromatics from lignin produces adipic acid and other acids plus diacids and polyols.

Bioproducts and Biochemicals From Syngas

In principle, syngas (primarily consisting of CO and H_2) can be produced from any hydrocarbon feedstock, including natural gas, naphtha, residual oil, petroleum coke, coal, and biomass. The extensive research and development devoted to syngas conversion to fuels and chemicals are documented in a vast literature that tracks the scientific and technological advancements in syngas chemistry.[13] Biomass gasification was introduced in Chapter 3 as an advanced biofuels process for converting clean, biomass-derived syngas into gasoline and diesel by the Fischer-Tropsch process. Bioproducts and biochemicals can also be produced if the syngas comes from a renewable carbon resource.

Hydrogen is separated from syngas and purified for fuel or used as a reagent in making ammonia and methanol. Haber-Bosch synthesis combines hydrogen and nitrogen over a catalyst at high pressure to produce ammonia. Hydrogen is also a common reagent for hydrogenation reactions to make a wide variety of chemicals. Catalytic methanol synthesis from syngas is a classic high-temperature, high-pressure, exothermic, equilibrium limited synthesis reaction with overall conversion efficiency of over 99%.

Methanol is an important intermediate in making many chemicals. For example, methanol dehydration over an acidic catalyst produces dimethyl ether, which is a diesel substitute, aerosol propellant, and precursor for olefin (ethylene, propylene, and butenes) and gasoline-range hydrocarbons. Formaldehyde can be made by catalytic partial oxidation of methanol with air over an unsupported Cu catalyst at atmospheric pressure. Formaldehyde is the main component for producing phenol-formaldehyde resins used in particle board and other construction products. The acid catalyzed reaction of isobutene and methanol produces methyl tertiary butyl ether, which was until recently an oxygenated additive in gasoline. The carbonylation reaction of methanol and CO produces acetic acid, a building block for vinyl acetate, acetic anhydride, and terephthalic acid, important reagents for making resin adhesives, polymers, and plastics.

Several other routes from syngas to products have been developed. A variant of methanol synthesis produces a mixture of higher alcohols (e.g., ethanol, butanol, and pentanol), but poor selectivity and low yield have limited commercial exploitation. Variations of Fischer-Tropsch synthesis have been developed to make other chemicals. The oxosynthesis process is the hydroformylation of olefins with syngas. It is the principal industrial chemical route to C_3–C_{15} aldehydes, which are converted into alcohols, acids, and other derivatives and used to make solvents, synthetic detergents, flavorings, perfumes, healthcare products, and other high value commodity chemicals.

Bioproducts and Biochemicals From Bio-oils

Thermochemical direct biomass liquefaction technologies (fast pyrolysis, catalytic fast pyrolysis, hydropyrolysis, and hydrothermal liquefaction) have focused on converting non-food biomass into liquids that compete with fossil fuel. For advanced biofuels production, oxygen removal improves the thermal stability of the hydrocarbon-rich liquid intermediates so that upgrading is less challenging and requires less hydrogen. However, there may be a missed opportunity if the oxygen-containing compounds that are being deoxygenated, removed, or destroyed have value as co-products. They can be recovered as chemicals prior to biofuel production to improve the carbon efficiency in the value chain, process efficiency, and ultimately the economic competitiveness of direct biomass liquefaction.

The liquid intermediates obtained from biomass pyrolysis contain useful molecules with unique oxygen functionalities that are uncommon in petroleum refining streams but have an advantaged use as raw materials or chemical building blocks in several sectors of the petrochemical industry. Unfortunately, the complexity and low concentration of individual compounds make recovery of marketable products very challenging.

Recovery of chemicals from direct biomass liquefaction products such as pyrolysis oils is not a new idea. Many research groups have explored options of extracting high-value chemicals from bio-oils[14] that are known to contain hundreds of oxygenated compounds[15] such as sugars, anhydrosugars, carboxylic acids, hydroxyaldehydes, hydroxyketones, and phenolics.

Catalytic fast pyrolysis readily depolymerizes biomass into a variety of chemical species. The product slate is controlled by the catalyst, process conditions, and the composition of the feedstock. Aromatic hydrocarbons such as benzene, toluene, and xylenes are made when zeolite catalysts are used.[16] Also, useful oxygenated compounds like ketones, furans, and phenols result when metal oxides like CeO_2 and TiO_2 are used.[17] Catalytic fast pyrolysis can also be tailored to make other oxygenated compounds such as levoglucosan, levoglucosenone, furfural, hydroxyacetaldehyde, acetic acid, gluconic acid,[18] and phenolic compounds.[19, 20]

Biochar

The goal of direct biomass liquefaction is to maximize liquid biocrude yield. Nevertheless, 15% or more of the feedstock is converted into a carbon-rich solid called char. This is typically burned in an integrated biomass pyrolysis process to provide heat to drive the endothermic devolatilization. If excess heat is available or if pyrolysis gases are recycled for energy or supplemental fuel is used, then there is potential to recover the char as a byproduct. It is often called biochar to distinguish it from charcoal, and there is rising awareness of the benefits of using it as a soil amendment.[21] It improves crop nutrient bioavailability, reduces nutrient leaching, increases water holding capacity, and improves aeration and porosity. Applying it to agricultural land and forest plantations is truly an example of carbon sequestration. Unfortunately, 1–10 tons of biochar per acre produce only modest agricultural gains, and the cost makes it uneconomical. It can be used in greenhouses for specialized applications such as fruit and vegetable production. Nevertheless, soil carbon sequestration and soil improvement by biochar is an ongoing area of research.

Other applications include filtration in water treatment and waste treatment facilities to remove organic impurities and heavy metals such as lead and nickel. It can also be "activated" for use in higher value applications that require novel carbon materials, such as lithium ion batteries, catalyst supports, and as an additive in steel manufacturing. These higher value applications require biochar with very specific chemical and material properties. Transforming it into high-value carbon is also an active area of research and development.

Bioproducts and Biochemicals From Microalgae

Algae are unique biomass materials that grow rapidly with high photosynthetic efficiency in aquatic environments. Microalgae differ from terrestrial biomass in that they do not contain lignin. The major components are carbohydrates, lipids, and proteins. Pretreatment processes developed for cellulosic ethanol production—with, for example, dilute acid—hydrolyze the carbohydrate fraction separated from algal biomass to produce monomeric sugars that are fermented to ethanol. The hexane solvent extraction used in the oleochemicals industry can be applied to recover algal oil. Hydrotreating algal oil produces renewable diesel blendstocks. The remaining algal meal is rich in protein, making it an excellent dietary supplement for fish, livestock, and poultry.

A model algal biorefinery would have three revenue streams: cellulosic ethanol, renewable diesel or jet fuel, and animal feed.[22] However, the high cost of algal biomass production for these high-volume, low-value commodities challenges the economic viability of an integrated algal biorefinery. Microalgae feedstock is well known for high-value, low-volume derivatives in the pharmaceutical and nutraceutical industries.[23] Higher value product options that increase revenue without upsetting existing markets could improve process economics for integrated algal biorefineries.

The chemical composition of the major components in microalgae is species specific. Like terrestrial biomass, the composition of the monomeric sugar stream from the carbohydrate fraction is a function of the species. So is the composition of algal oils (triglycerides, phospholipids, sulfolipids, and free fatty acids). The protein chemistry and available pigments also depend on the type of microalgae. The composition of algal biomass is also highly dynamic and changes in response to growing conditions (nutrient loading, light intensity, etc.). For example, nutrient deprivation leads to high lipid production in select species. Therefore, the biodiversity of microalgae and the dynamic accumulation of major components can be very advantageous for identifying and optimizing the production of higher-value products.

Carbohydrates separated from microalgae can be hydrolyzed to monomeric sugars to produce the building block chemicals and subsequent biobased chemicals, plastics, and polymers discussed above that have higher value than ethanol. Higher-value products from algal oils include oleochemicals used in cosmetics, nutraceuticals, paints, and surfactants. Carotenoids such as lutein and β-carotene and polyunsaturated fatty acids can be recovered and sold as dietary supplements. The amino acid profile is important for determining the quality of the remaining protein fraction as a nutritional supplement or animal feed.

The potential market applications for chemical feedstocks from microalgae covers the gamut of products from plastics, surfactants, plasticizers, and polyurethane to nutraceuticals and protein for feed and food applications. Many of these products are currently produced from fossil fuel feedstocks, often under hazardous reaction conditions using toxic reagents and solvents that have a high CO_2 footprint. Switching to production pathways from microalgae could hasten the transition to more sustainable chemicals and materials.

Economic and Market Considerations

So, what products should we make from biomass? Ideal candidates would have high market value and volume and an advantaged synthesis route where the product could be made from biomass at a lower cost than from a fossil fuel feedstock. Fig. 11.3 shows potentially attractive candidates for bioproducts based on market volume vs market value. Gasoline and diesel are also shown for reference, to highlight that all chemical products are significantly lower in market volume than fuels, but many have relatively high volume with higher market price than fuels.

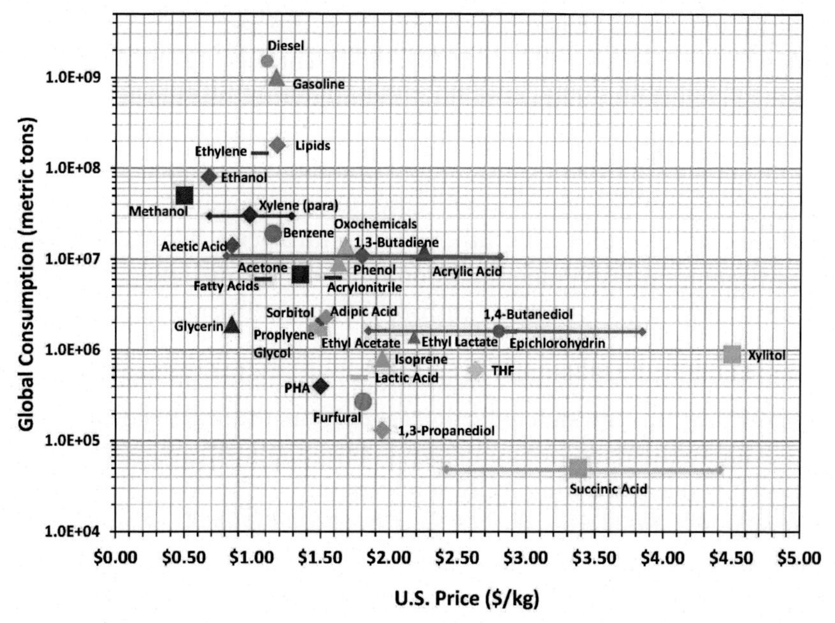

Fig. 11.3 Bioproducts and market opportunities. *(Recreated from Biddy MJ, Scarlata C, Kinchin C. Chemicals From Biomass: A Market Assessment of Bioproducts With Near-Term Potential. National Renewable Energy Laboratory; 2016. March 2016. NREL/TP-5100-65509.)*

But it is also important to consider other commodity prices and market dynamics. Take methanol, for example. As described earlier, methanol and its derivatives are readily made from biomass via gasification and syngas conversion, but competing in the open market with methanol made from natural gas will be economically challenging. The US Energy Information Administration projects low-cost natural gas for the foreseeable future, with considerable amounts of new supply coming on the market in the next decade. It would be difficult for methanol from biomass to gain market share unless the production cost from natural gas could be matched.

However, chemical markets are dynamic and interrelated, so low-cost natural gas could also open up potential market opportunities for bio-based chemicals. For example, C_4 and C_5 chemicals such as isoprene and butadiene are typical byproducts when crude oil is used as the feedstock for ethylene production. Now that natural gas has become the preferred feedstock for ethylene production, the isoprene and butadiene markets will be undersupplied. This represents an opportunity for biomass-based processes.

Another opportunity space for bio-based chemicals is relatively modest markets, compared to fuels and high-volume commodity chemicals, that could be significantly increased if lower cost processes were developed. A good example is acrylonitrile (ACN), a starting material in many production processes such as those for ABS plastics, carpet fibers, and carbon fiber. Roughly 40% lighter and 10 times stronger than aluminum, composite materials made with carbon fibers have the potential to reduce the weight of automobiles and improve fuel efficiency to well over 50 mpg. However, the current process for ACN synthesis starts with a petrochemical feedstock and involves a propylene ammoxidation reaction that generates a substantial amount of highly toxic hydrogen cyanide and has a very large environmental footprint. Also, it takes almost 2 kg of ACN to make 1 kg of carbon fiber, and large amounts of CO_2 are released in the manufacturing process. All of the greenhouse gas emission reduction gained from the improved fuel economy of lightweight vehicles using carbon fiber composites would be offset by the CO_2 released during ACN production.

Recently, a group from the National Renewal Energy Laboratory demonstrated ACN made from lignocellulosic biomass.[24] ACN was produced at molar yields exceeding 90% from 3-hydroxypropionic acid, which can be made microbially from sugars. The endothermic approach described eliminates the threat of runaway reaction hazards that occur with ammoxidation, and it avoids the creation of hydrogen cyanide while greatly reducing the CO_2 footprint compared to ACN production from petrochemicals.

When considering options for bio-based products and chemicals, it is important to understand market opportunities that match the scale (volume) and price points for co-product streams in an integrated biorefinery. The co-products may be identical to existing chemicals, fuels, and products, or identical in functional performance but otherwise advantaged, or new materials with advantaged performance characteristics. Price parity is expected for drop-in replacements. Therefore, additional technology development should focus on reducing production cost. The pathway to market acceptance is fairly streamlined for direct replacements if the new chemical constituency and purity can be proven to be equal to or greater than those of the existing chemical. The pathway to market acceptance for functional replacements is more complicated, because extensive product testing will be required to prove that the new product has equivalent or better performance than the one made from the chemical being functionally replaced.

New materials are the most challenging, as new markets need to be developed and price points established. For new novel bio-based chemicals, the pathway to market acceptance will be extensive. The onus will be on the producer to prove that the new product has value in the marketplace to justify the cost and risk associated with introducing it. Bio-based products and chemicals that have similar functional performance to existing ones are expected to be cost competitive but may have commercial advantage for being bio-derived.

Consider vanillin as an example. It can be recovered from vanilla beans and marketed as pure vanilla extract, commanding a high price. It can be synthesized in a multi-step chemical process starting with petrochemical precursors to produce imitation vanilla, sold at a much lower price. Clove oil can be extracted to isolate guaiacols and eugenols that can be used to synthesize vanillin in a much simpler two-step process. The starting material is an oil extracted from biomass, so technically it is a bio-derived product. Consequently, it sells at an intermediate price.

The primary challenge in maximizing the profitability of an integrated biorefinery is to match the scale of high-value, low-volume bio-products with the scale of high-volume, low-value commodity biofuels without disrupting the markets. The maximum biofuels potential in the United States is roughly one-quarter to one-third of the total transportation fuel market, based on the availability of sustainable biomass feedstocks. The markets for many high-value bio-based products and chemicals are much lower volume. Market research to estimate future demand for specific products is absolutely necessary to insure long-term profitability of future integrated biorefineries. The financial risk associated with developing new technology for biofuels is high. Co-producing high value bioproducts and biochemicals is a valid option for improving process economics to reduce this risk. Careful selection of products is necessary to avoid adversely affecting or saturating certain markets as the bioeconomy continues to develop.

Close-up: Vanillin From Biomass

Vanillin, a phenolic aldehyde, is an important specialty chemical with many applications in the flavor and fragrance industry. In 2016, the annual global market was about 18,650 tons.[25] Currently, only 1% of the total market share is made up of natural vanillin; the remaining 99% is via synthetic routes.[26, 27] Synthetic vanillin can be made by several chemical methods as well as biosynthesis and biotechnological routes with starting materials such as coniferin, guaiacol, eugenol, lignin, ferulic acid, and even glucose, as illustrated in Fig. 11.4.

The synthetic vanillin made from petroleum-derived guaiacol accounts for about 85% of the market share. The route to guaiacol and then to vanillin involves several major steps, as illustrated in Fig. 11.5 (1): conversion of benzene to cumene; (2) oxidation of cumene to cumene hydroperoxide; (3) decomposition of the hydroperoxide with sulfuric acid to make phenol,

Continued

Close-up: Vanillin From Biomass—cont'd

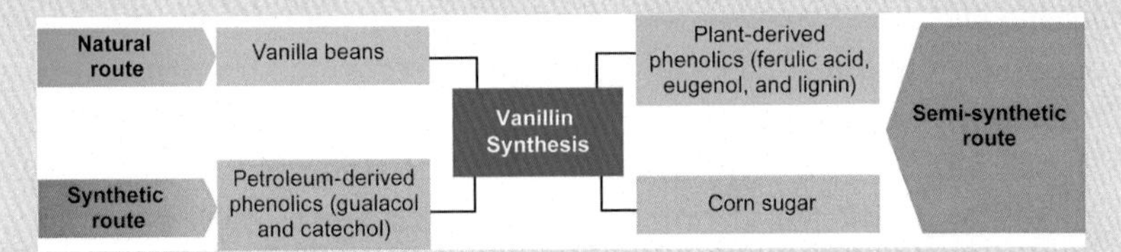

Fig. 11.4 Several synthesis routes to vanillin.

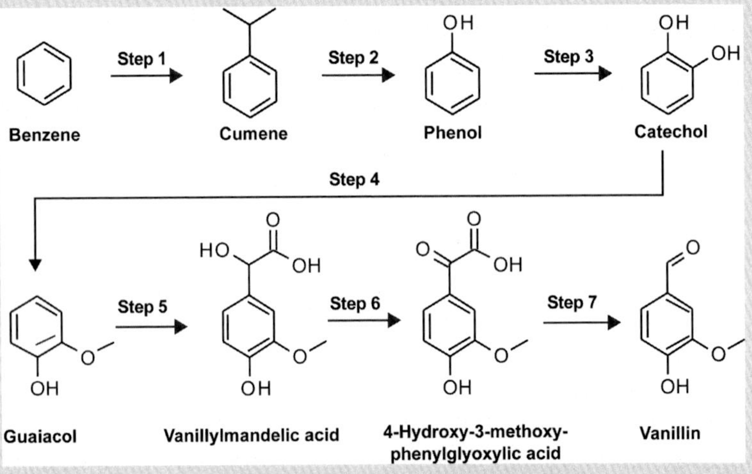

Fig. 11.5 Petroleum route to vanillin starting from benzene.

which is then hydroxylated with peroxide to catechol; followed by (4) monomethylation of the catechol using methyl halides to make guaiacol. The guaiacol is (5) condensed with glyoxylic acid in an alkaline media to make mandelic acid, followed by (6) oxidation to the corresponding phenylglyoxylic acid that (7) decarboxylates to produce vanillin. This is just one of many cumbersome ways oxygenated chemicals are synthesized from petroleum feedstocks. It underscores the fact that biomass feedstocks have an advantage when it comes to the production of oxygenated products.

Routes that use non-petroleum, environmentally friendly reagents are in large demand. RTI recently demonstrated the proof of concept for using the eugenols and isoeugenols present in biocrude produced by catalytic fast pyrolysis of loblolly pine. It is a two-step approach, as illustrated in Fig. 11.6. First the eugenols are isomerized to increase the isoeugenol concentration; then the isoeugenols are selectively oxidized to vanillin. One of the most efficient isoeugenol oxidation reagents for vanillin production is ozone, a

Close-up: Vanillin From Biomass—cont'd

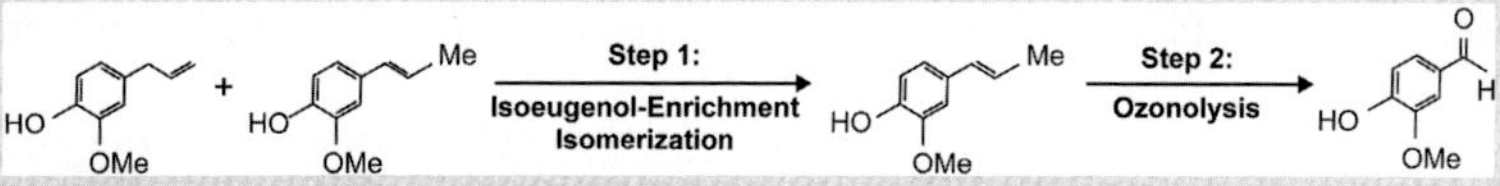

Fig. 11.6 Two-step synthesis of vanillin from biocrude components developed at RTI.

green oxidant that decomposes into molecular oxygen and can be produced at a large scale without environmental concern. The efficient production of high-purity vanillin via targeted oxidation of biomass-derived eugenols and isoeugenols provides a more environmentally safe process than the multistep synthesis route starting with petroleum intermediates.

Ofei D. Mante
RTI International, Research Triangle Park, NC, United States

References

1. Rong G, Zhang Y, Zhang J, Liao Z, Zhao H. A robust engineering strategy for scheduling optimization of refinery fuel gas system. *Ind Eng Chem Res*. 2018;57:1547–1559.
2. Vennestrøm PNR, Osmundsen CM, Christensen CH, Taarning E. Beyond petrochemicals: the renewable chemicals industry. *Angew Chem Int Ed*. 2011;50(45):10502–10509.
3. Luterbacher JS, Martin Alonso D, Dumesic JA. Targeted chemical upgrading of lignocellulosic biomass to platform molecules. *Green Chem*. 2014;16(12):4816–4838.
4. Sheldon RA. Green and sustainable manufacture of chemicals from biomass: state of the art. *Green Chem*. 2014;16(3):950–963.
5. Walsh P, de Jong E, Higson A, Wellisch M. *Bio-Based Chemicals: Value Added Products From Biorefineries*. http://wwwiea-bioenergytask42-biorefineriescom/ (May 2012); 2012.
6. Turley DB, Chaudhry Q, Watkins RW, Clark JH, Deswarte FEI. Chemical products from temperate forest tree species—developing strategies for exploitation. *Ind Crop Prod*. 2006;24(3):238–243.
7. ASTM. Standard test methods for testing tall oil. In: *Subcommittee D01.34 on Pine Chemicals and Hydrocarbon Resins*. West Conshohocken, PA: ASTM International; 2012. vol. ASTM D803-03.
8. Drew J, Russell J, Bajak HW. In: Drew J, et al., eds. *Sulfate Turpentine Recovery*. 1971.
9. Zinkel DF, Russell J. *Naval Stores: Production, Chemristry, Utilization*. Pulp Chemicals Association; 1989.
10. Werpy T, Petersen G. *Top Value Added Chemicals From Biomass: Volume I—Results of Screening for Potential Candidates From Sugars and Synthesis Gas*. 2004. August NREL/TP-510-35523.
11. Isikgor FH, Becer CR. Lignocellulosic biomass: a sustainable platform for the production of bio-based chemicals and polymers. *Polym Chem*. 2015;6(25):4497–4559.
12. Ragauskas AJ, Beckham GT, Biddy MJ, et al. Lignin valorization: improving lignin processing in the biorefinery. *Science*. 2014;344(6185)1246843.

13. Spath PL, Dayton DC. *Preliminary Screening—Technical and Economic Assessment of Synthesis Gas to Fuels and Chemicals With Emphasis on the Potential for Biomass-Derived Syngas*. Golden, CO: National Renewable Energy Laboratory; 2003. TP-510-34929.
14. Bridgwater AV. Production of high grade fuels and chemicals from catalytic pyrolysis of biomass. *Catal Today*. 1996;29(1–4):285–295.
15. Czernik S, Bridgwater AV. Overview of applications of biomass fast pyrolysis oil. *Energy Fuel*. 2004;18(2):590–598.
16. Carlson T, Tompsett G, Conner W, Huber G. Aromatic production from catalytic fast pyrolysis of biomass-derived feedstocks. *Top Catal*. 2009;52(3):241–252.
17. Mante OD, Rodriguez JA, Senanayake SD, Babu SP. Catalytic conversion of biomass pyrolysis vapors into hydrocarbon fuel precursors. *Green Chem*. 2015;17(4):2362–2368.
18. Kim J-S. Production, separation and applications of phenolic-rich bio-oil—a review. *Bioresour Technol*. 2015;178:90–98.
19. Zhu X-f, Lu Q. Selective fast pyrolysis of biomass to produce fuels and chemicals. In: Lee JW, ed. *Advanced Biofuels and Bioproducts*. New York: Springer; 2013:129–146.
20. Santhanaraj D, Rover MR, Resasco DE, Brown RC, Crossley S. Gluconic acid from biomass fast pyrolysis oils: specialty chemicals from the thermochemical conversion of biomass. *ChemSusChem*. 2014;7(11):3132–3137.
21. Schmidt HP, Wilson K. The 55 uses of biochar. *Biochar J*. 2014. https://www.biochar-journal.org/en/ct/2. Accessed 11 May 2019.
22. Laurens LML, Markham J, Templeton DW, et al. Development of algae biorefinery concepts for biofuels and bioproducts; a perspective on process-compatible products and their impact on cost-reduction. *Energy Environ Sci*. 2017;10(8):1716–1738.
23. Bhalamurugan GL, Valerie O, Mark L. Valuable bioproducts obtained from microalgal biomass and their commercial applications: a review. *Environ Eng Res*. 2018;23(3): 229–241.
24. Karp EM, Eaton TR, Sànchez i Nogué V, et al. Renewable acrylonitrile production. *Science*. 2017;358(6368):1307–1310.
25. Grand View Research. *Vanillin market research report*. 2017. GVR-1-68038-836-7.
26. Fache M, Boutevin B, Caillol S. Vanillin production from lignin and its use as a renewable chemical. *ACS Sustain Chem Eng*. 2016;4(1):35–46.
27. Gallage Nethaji J, Møller Birger L. Vanillin–bioconversion and bioengineering of the most popular plant flavor and its de novo biosynthesis in the vanilla orchid. *Mol Plant*. 2015;8(1):40–57.

Additional Reading

Beckham GT, ed. *Lignin Valorization: Emerging Approaches*. The Royal Society of Chemistry; 2018.

Biddy MJ, Scarlata C, Kinchin C. *Chemicals From Biomass: A Market Assessment of Bioproducts With Near-Term Potential*. National Renewable Energy Laboratory; 2016. March 2016. NREL/TP-5100-65509.

Bisaria VS, Kondo A. *Bioprocessing of Renewable Resources to Commodity Bioproducts*. John Wiley & Sons, Inc.; 2014.

Gallezot P. Conversion of biomass to selected chemical products. *Chem Soc Rev*. 2012;41(4): 1538–1558.

Isikgor FH, Becer CR. Lignocellulosic biomass: a sustainable platform for the production of bio-based chemicals and polymers. *Polym Chem*. 2015;6(25):4497–4559.

Lee JW. *Advanced Biofuels and Bioproducts*. New York: SpringerLink; 2013.

Snyder SW. *Commercializing Biobased Products: Opportunities, Challenges, Benefits, and Risks*. Royal Society of Chemistry; 2016.

CHAPTER TWELVE

Waste to Energy

Introduction

Since the dawn of civilization, humankind has been faced with the concept of waste management as the formation of groups and communities concentrated consumption and utilization of raw materials for sustaining life into relatively small areas. Collection of raw materials, food, and water was challenging and often done at great peril, so whatever was collected was fully used with little waste. Consumption of raw materials and energy has increased with growing population and the higher standard of living afforded by industrialization and modernization. So has the production of waste that needs to be remediated. Population growth in urban areas during and after the industrial revolution led to increasing waste production with no real intention to manage it. Uncontrolled open dumping in rivers, on the streets, and outside of town led to polluted air, water, and soil that became breeding grounds for bacteria and disease. These unsanitary conditions caused serious health problems for urban populations. Poor environmental quality became synonymous with a decrease in quality of life, and public uproar quickly led to the development of waste management strategies.

As urbanization increases, prosperity improves the quality of life and increases disposable income. Greater wealth then leads to greater consumption of goods and services that increase the amount of waste generated. The global economy is driven by consumption, and the byproduct of economic development and prosperity is waste production. The volume of waste continues to increase as population surges, demanding continuously evolving strategies for waste management. Based on a recent study published by the World Bank,[1] nearly 1.3 billion tonnes of municipal solid waste (MSW) were generated globally in 2012, with the expectation that this will nearly double to 2.2 billion tonnes per year by 2025.

The composition of waste is also continuously changing. Technological advancements have led to new product choices, and as per capita wealth increases, consumers can afford to take advantage of this variety. In the early

Analytical Methods for Biomass Characterization and Conversion
https://doi.org/10.1016/B978-0-12-815605-6.00012-3

20th century, the primary raw materials were wood, metal (iron ore), and glass. As a result, inorganic materials were the majority component of municipal waste. The growth of the petrochemical industry and the discovery of new materials and processes led to an increase in the production of plastics for packaging and of non-ferrous metals like aluminum as a lightweight alternative to steel. Single use materials now make up the bulk of municipal waste, and inorganic materials are separated out for recycle and reuse.

Developing strategies for managing this enormous volume of waste require cost effective solutions that are socially acceptable and environmentally sustainable. Landfills were created as a low-cost option for managing MSW; however, sprawling population centers created competition for land close to urban areas and environmental and health issues as communities were developed in proximity to landfills. Eventually, the capacity of landfills is reached or exceeded, and new sites are needed. Transportation costs can become a burden as these sites get further from the source.

Incineration has been used for over 100 years to reduce MSW volume. The first incinerator in the United States was commissioned in 1885 on Governors Island in New York City.[2] Early incinerators were built without interest in energy recovery. Incinerator technology soon took advantage of the thermal energy in MSW by recovering it for heat and steam production. However, the cost of energy recovery was driven up by the increasing cost of environmental controls for air emissions reduction. Public concerns about pollution from MSW incinerators forced these facilities further away from population centers, requiring more infrastructure (and cost) to deliver the energy to consumers. Public opposition to MSW incineration has made it less attractive.

The most obvious strategy for minimizing waste is to produce less of it. The close link between economic growth and increased waste is well known, but reducing economic activity is not a viable option. Therefore, new paradigms for resource management are needed to maintain prosperity with environmental sustainability. The concept of the circular economy is based on the premise that raw materials used to produce goods remain in the supply chain for as long as possible to minimize resource consumption and waste. This requires that products be designed for reuse, not single use and disposal. Energy efficient, cost effective recycling returns the materials to the front of the supply chain, reducing raw material consumption. As products approach end of life, their energy content can be recovered through a variety of waste to energy processes that reduce net greenhouse gas emissions by offsetting consumption of additional fossil fuels. The circular economy concept goes beyond just recycling and combined heat and power production. It starts with the vision for maximizing the use of raw materials by designing products for use, reuse, recycling, and ultimately energy recovery. The following sections briefly touch on examples related to these issues in the context of waste to energy.

Waste Management

Waste management is the complex integration of resource management, economic development, public policy and acceptance, and government regulations while protecting the environment, preserving economic prosperity, and improving quality of life. Waste management technology options include physical separation and recovery, thermal conversion (combustion, gasification, and pyrolysis), chemical transformation, and biological conversion (aerobic and anaerobic). Strategies for waste management are highly region specific. Waste volumes, technology options, land availability, local infrastructure, social acceptance, and environmental regulations all affect decisions, but the primary driver that links them is cost. The best solutions for a given municipality, city, or region maximize the overlap between social, environmental, and regulatory considerations while minimizing cost. After all, the idea of throwing something away is a fallacy because waste always goes somewhere in some form whether it is buried underground or released as emissions when burned. Therefore, maximizing the utilization of raw materials entails reducing consumption and reusing and recycling as much as possible before energy recovery and ultimately landfill disposal.

Recycling

MSW is separated into organic and inorganic fractions. The organic fraction includes food and yard waste that can be used for energy or transportation fuel production. Other components such as paper, plastic, glass, and ferrous and non-ferrous metals make up the inorganic fraction.

An important part of the waste management value chain is the materials recovery facility (MRF). These specialized units are designed to receive municipal waste, separate out recyclable materials, and provide these materials for secondary raw material end-users. The degree to which MSW has been segregated defines the needs of the MRF. A so-called dirty MRF processes a mixed waste stream that contains both the organic and inorganic (whole) fractions. Manual and mechanical sorting recovers recyclable materials, and the remainder is incinerated for energy recovery or disposed of in a landfill. This type of facility is generally located in regions where separating organic and inorganic streams at the source (curbside or community recycling center) is ineffective or too costly. Clean MRFs process the segregated inorganic components for sale into the market. Efficient, economical source separation of MSW requires the infrastructure for collecting multiple, segregated waste streams and maximum public participation.

Separated paper and cardboard is reprocessed into new paper products. The first step is to remove the ink and separate the cellulose fibers from the added fillers. Once the cellulose fibers are recovered as pulp, they are used to make recycled paper. Producing pulp from wood requires significant heat and chemical energy to dissolve lignin and recover cellulose fibers. Recycled paper contains no lignin, so it requires less energy to produce. However, there is a limit to the number of times cellulose fibers can be recycled. Eventually they become too short, requiring addition of new longer ones to maintain product quality standards.

Ferrous metals such as steel can easily be captured by magnets and sold as scrap to be smelted and recast. Non-ferrous metals include aluminum, copper, and brass. Copper has high value and is often source segregated. Aluminum, mostly beverage cans, is separated, recovered, and remelted. Remelting aluminum is much less energy intensive than producing it from bauxite (aluminum oxide).

Post-consumer plastics make up much of the segregated inorganic fraction of MSW. Plastics are ubiquitous for packaging, beverage containers, laminates, building materials, automobiles, and many other consumer products. Within this myriad of end-use applications, there are seven categories (Table 12.1), based on specific physical and chemical properties recognized globally. Identification numbers help consumers sort plastics for recycling. However, uses for mixed plastic streams are developing, and sorting

Table 12.1 The Seven Categories of Plastic and Their Identification Numbers.

1. Polyethylene terephthalate (PETE or PET)
2. High-density polyethylene (HDPE)
3. Polyvinyl chloride (PVC)
4. Low-density polyethylene (LDPE)
5. Polypropylene (PP)
6. Polystyrene or Styrofoam (PS)
7. Miscellaneous: polycarbonate, polylactide, acrylic, acrylonitrile butadiene, styrene, fiberglass, nylon

processes that capitalize on physical property differences (mainly density and melting point) are evolving. Optical methods that can sense the chemical differences between plastics are also being implemented.

Chemical recycling of plastic waste is the most noteworthy polymer recovery technique for producing monomers that can be reused to make the same plastic polymers. Monomer recovery reduces the consumption of fossil resources (petroleum and natural gas) that are initially used for producing plastics. Recovering monomers and other chemical precursors from waste plastics is hampered by additives (dyes, colorants, and copolymers) and impurities (plastic and paper labels, adhesives, and inks) that need to be removed to maintain the required purity of the starting materials.

Wastewater Treatment

There are 15,014 publicly owned treatment works (POTWs) in the United States.[3] Historically, national estimates of municipal sludge and treated biosolids are based on biosolids production, because managing treated sludge as MSW requires an understanding of the quantity of residual material to be disposed of rather than the amount of sludge initially produced. In the absence of annual sludge production data for POTWs in the United States, a simplified model was used by Seiple and colleagues to estimate the amounts of primary, secondary, and total sludge, on a dry matter basis, as a function of the amount of wastewater treated.[4] Those 15,014 POTWs treat 34.5 billion gallons of wastewater per day, generating 13.84 million tons of sludge per year.[4] For the purpose of assessing potential biofuel conversion, POTWs produce 11.76 million tons per year of sludge on an ash free dry weight basis.

Suitable technology options for sludge and biosolids handling, use, and disposal are scale dependent. Small facilities treat <1.5 million gallons per day (MGD) of wastewater and produce <1.5 dry tons per day (DTPD) of sewage sludge. They account for 79% of POTWs (about 11,000 plants) and treat only 1% of the total wastewater flow. Most of these plants do not have an anaerobic digester onsite; therefore, sludge disposal is an added cost. The sludge is usually dried to between 8% and 10% moisture to minimize transportation costs for land application or landfill disposal.

Wastewater flow in medium-size plants is between 1.5 and 15 MGD and produces 1.5–15 DTPD of sludge. They account for 18% of POTWs (about 2700 plants) and treat 30% of the total wastewater flow.

For large plants, wastewater flow is between 15 and 150 MGD and produces 15–150 DTPD. They account for 3% of POTWs (414 plants) and treat 39% of the total wastewater flow. Super large plants treat more than 150 MGD and produce more than 150 DTPD, but there are only 31 of them.

About 1200 medium to large plants in the United States use anaerobic digesters. They treat about 18.2 billion gallons per day of the inflow and process 7.3 million dry tons of sludge per year. This amount is 53% of the total sewage sludge generated by the POTWs, suggesting that just under half of the total amount of sewage sludge is underutilized.[4] This is a significant resource for biofuels and bioproducts.

The average cost of disposal for biosolids is regionally dependent and varies between $30 and $100 per dry ton. Disposal costs are high in New England and densely populated areas, low in Montana, Kentucky, and West Virginia. Therefore, site selection has a large role in the economics of a future waste to energy facility.

Waste to Energy Technology Options

Feedstocks

There are many categories of waste: agricultural residues beyond grain harvesting (sugarcane bagasse, cotton gin trash, orchard prunings, pits and shells, soybean and rice hulls); fats, oils, and grease (FOG—poultry and livestock fats, used cooking oil, yellow grease, brown grease); MSW (also called household waste, urban waste, garbage, trash); construction and demolition waste; medical waste; electronic waste; industrial waste; food waste (residential, commercial, institutional, industrial); biosolids from wastewater treatment; and animal manures (swine and dairy, poultry litter).[5] Not including FOG and MSW, the 2016 update to the Billion Ton Study estimates that there is the potential for 34 million dry tons of agricultural waste feedstocks and animal manures available in the United States annually.[6] By far the largest potential for waste to energy is from landfill gas recovery for biopower generation or conversion to renewable natural gas.

The complexity and variability of waste feedstocks presents a challenge for waste to energy processes. Moisture is the component that varies most widely, followed by organic composition and ash content. Technically, the amount of moisture in a feedstock limits the conversion pathway for producing biofuels, bioproducts, and biopower. Thermochemical conversion technologies are considered for dry feedstocks (<50% moisture) that have well-understood material and preparation costs. Waste feedstocks can have higher moisture content but are attractive for dedicated waste to energy conversion because they are usually delivered and concentrated at specific locations, but the energy penalty for dewatering and drying is cost prohibitive.

A cost effective, energy efficient dewatering and drying process is essential for adding value to high-moisture streams.

Incineration

Incineration has been used since the mid-19th century to reduce the volume of waste generated in urban areas. It wasn't until the mid-20th century that energy recovery from waste incineration was practiced. The goal of MSW incineration is threefold: reduce waste volume and recover energy and materials while improving environmental performance with stringent emissions control.

The composition of MSW is extremely variable. The energy content is directly affected by the amount of combustible material. Consequently, presorting to remove inert components (metals, glass, construction materials, etc.) increases the calorific value. But another major factor is the moisture content of the sorted material. Variation in energy content affects heating rate, combustion efficiency, and emissions control, and can lead to reduced capacity from frequent unscheduled process interruptions.

Incinerators are based on three types of combustor design: moving grate, rotary kiln, and fluidized bed. Moving grate systems are much more forgiving in terms of the size and composition of the input MSW. They can accommodate large volumes with minimal material pretreatment (sorting and shredding) and are not very sensitive to variations in composition, particularly energy content dictated by moisture content and inert materials. Consequently, they are more expensive than the other two. While the capital costs for fluidized bed systems are lower, they are more difficult to operate because of sensitivity to feedstock homogeneity. And they require the most MSW preprocessing, mainly shredding and size classification, to maintain an even energy content fed into the combustor. Inert materials can build up in the bed and agglomerate, causing unstable operation and ultimately interruption from bed defluidization.

Energy is recovered from MSW incineration as hot water or steam using low pressure or high pressure boilers or water wall furnaces. Designing for optimal heat exchange is done within the context of the different combustor designs. Heat recovery for electricity requires maximum combustion efficiency to produce high quality steam. Combined heat and power production can be done to maximize thermal energy recovery, but the location of the plant determines the end use. For example, lower grade thermal energy can provide hot water for residential or industrial heating if the infrastructure exists, and waste heat can be captured for sewage sludge drying if the plant is co-located with a wastewater treatment facility.

The energy recovery potential and volume reduction benefits of waste to energy facilities is often not enough to counteract the negative public opinion about environmental concerns regarding air emissions and residual solid waste (ash) handling. Early incinerators had little if any air pollution control equipment. As a result, uncontrolled emissions of nitrogen and sulfur oxides caused acid rain, combustion of PVC and other fuels with high chlorine content formed dioxins, and poor waste segregation caused heavy metal emissions, mainly lead, mercury, and cadmium. Public acceptance was low in the 1970s because of the stigma caused by the air pollution. Consequently, advancements in pollution control became critical for the industry.

Pollution control strategies involve both upstream and downstream solutions. Better waste segregation to remove hazardous materials (heavy metals, PVC plastics, paints, solvents, halogen lightbulbs, etc.) before they enter the combustor minimizes toxic air emissions. Careful control of the combustion process (temperature, excess air ratio and flue gas recirculation) mitigates emissions. But a comprehensive integrated control system is still essential for cleaning flue gas to meet strict emission requirements before release into the atmosphere.

Wet and dry scrubbing systems and selective and non-selective catalytic reduction systems are common air emission control methods. Particulate matter (fly ash) containing heavy metals and other inorganic ash components is removed from flue gas by electrostatic precipitators or baghouse filters. Wet scrubbing mitigates acid gas (HCl, SO_2, NO, NO_2, HF) at the risk of corroding equipment surfaces, and with the added expense of treating the wastewater effluent. Adding chemicals such as calcium carbonate (limestone) and sodium bicarbonate to scrubbing liquids helps neutralize the wastewater and precipitate out sulfides and chlorides for separation, recovery, and disposal. Selective catalytic reduction of nitrogen oxides is achieved by injecting ammonia or urea in the flue gas stream, sulfur oxides are controlled with limestone injection, and activated carbon is added to capture mercury and dioxins. A multi-stage, multi-technology approach is clearly needed, but the cost of the control equipment to meet strict air emission standards remains a challenge. Unfortunately, if the temperature is below the dewpoint of water in the flue gas, a thick white plume emanates from the stack, reinforcing the environmental stigma even if the system is compliant with regulations.

Gasification

The challenges associated with biomass gasification discussed in Chapter 3 are magnified by the variability of the composition (moisture and inorganic content) of waste feedstocks and by the potential for higher concentrations of impurities such as sulfur, nitrogen, chlorine, alkali metals, and heavy metals.[7] The high inorganic content of many waste feedstocks, compared to woody biomass, is an issue that affects the primary conversion step and the gasifier design. For example, carefully controlling temperature in fluidized bed gasifiers (600–850 °C) to avoid ash softening temperatures is important for managing bed agglomeration. Carbon conversion is adversely affected if gasification temperatures are too low, so an alternative strategy is to operate waste gasification above the ash melting temperatures to produce a molten slag that is removed and recovered. The energy content of the waste feedstocks ultimately limits the gasification temperature. Consuming all the carbon to maximize temperature comes at the expense of the syngas yield unless an external energy source is used in the gasification process. This is the basis for the plasma gasification process that uses plasma torches to create an electric arc that produces a very high temperature (10,000+°C) plasma gas. Temperature in a plasma gasifier is controlled by the amount of supplied external energy so that variations in feedstock energy content do not adversely affect the gasification process.

Selection of the gasification agent (oxidizer) is based on the desired end use of the syngas. Air or enriched air is used to produce a low to medium energy content syngas for heat and power production. Steam waste gasification avoids excess nitrogen in the syngas, making it suitable for catalytic upgrading to fuels and chemicals, as described in Chapter 3.

Equivalence ratio (fuel:oxider) and temperature are adjusted to vary the syngas composition. The typical equivalence ratio is around 0.3. Anything lower reflects less partial oxidation and more pyrolysis, and anything higher enhances oxidation until full combustion is reached at equivalence ratio 1.0. Lower gasification temperature reduces char conversion and increases CO_2 production relative to CO. Tar concentrations are also higher at lower gasification temperatures, reducing carbon efficiency and increasing operational difficulty. Methane and tar decomposition increase as gasification temperature rises above 1000 °C, and the higher temperatures that are achieved in a plasma gasifier produce molten slag as metals and ash components liquify. Even at plasma gasification temperatures, tars are not completely destroyed, and char is not completely converted to syngas.

Trace elements such as sulfur, nitrogen, and chlorine need to be scrubbed from the syngas before it is converted to electricity or fuel. H_2S, NH_3, and HCl leads to excessive corrosion and air pollutants if not removed before syngas is burned in an engine or turbine. These trace impurities are also known poisons in catalytic fuel synthesis processes.

Despite these challenges, waste gasification has several advantages over incineration. The syngas intermediate can be used for biopower production at higher efficiency in engines and gas turbines compared to the conventional Rankine steam cycle. Engine generators can also be implemented for power production at smaller scale, making gasification a more modular technology than incineration. Syngas can also be used to produce fuels and chemicals if stringent cleanup and conditioning specifications are met. Of course, economies of scale are lost in modular systems, so the economic viability of gasification needs to be carefully considered.

Pyrolysis

As another thermochemical conversion option for waste management, pyrolysis is best suited for dry feedstocks to maintain process thermal efficiency. Production of a liquid intermediate makes pyrolysis favorable for distributed processing and energy densification that matches the scale of waste generation and collection. The heterogenous nature of waste feedstocks yields a lower quality bio-oil that can be challenging for renewable energy production. As pointed out, waste feedstocks tend to have higher concentrations of heteroatoms (sulfur, nitrogen, and chlorine) and ash that end up as impurities in the liquid bio-oil and solid bio-char product streams. Consequently, MSW pyrolysis has not gained a lot of attention for waste management and renewable energy production, but pyrolysis of specific targeted waste streams has been successfully implemented for materials and energy recovery.

Wood wastes, such as construction and demolition debris, sawmill residues, waste pallets, hog fuel (bark), and forest residues are similar enough to conventional woody biomass feedstocks so that pyrolysis can be done at a scale appropriate for the quantity of available resources. Scrap tires, on the other hand, are an abundant resource that poses a unique challenge. Tires are designed to be thermally stable and as a result have very low volatile matter content. They have high inert content because of the steel structural reinforcement, and the sulfur content is quite a bit higher than in biomass because of the vulcanization process. Much preprocessing is required. Tires are first shredded to recover the metals and reduce the particle size to maximize heat transfer in the pyrolysis reactor. The primary process streams are oil, carbon black (char), and gas. Pyrolysis gases are burned for process energy, the oil is used as boiler fuel for energy production, and carbon black is sold as a byproduct.

Mixed waste plastics recovered from sorting municipal waste are another pyrolysis feedstock. Plastics pyrolysis yields a liquid oil that can be used as a boiler fuel for heat and power production or as a diesel or turbine fuel. It is rich in hydrocarbons, since plastics are produced from petroleum, so little or no upgrading of the liquids is required. It is critical, however, that PVC be completely removed from the mixed plastics stream to avoid oil with high chlorine content that would accelerate corrosion downstream.

Recovering monomers in the liquid oils from plastics provide feedstocks for new plastic production. Separating the mixed plastic stream into individual polymeric materials at the front end could be avoided if the resulting monomers were separated downstream of the pyrolysis process. Instead of separating high-density polyethylene, polystyrene, polypropylene, and polyethylene terephthalate, the mixed plastic stream could be pyrolyzed and ethylene and propylene recovered from the gas phase. Styrene and terephthalic acid could be recovered in the liquid and then separated into monomer products. Of course, any other impurities in the feedstock, such as paper labels, adhesives, and other plastics, can be separated as well.

Biological Conversion

Biological conversion can occur aerobically or anaerobically. Aerobic processes are much faster, but energy recovery is not practical, as most organic carbon is converted to CO_2 and heat. Aerobic conversion is composting, and composting requires biodegradable carbon, nutrients (like nitrogen), oxygen, and water. As organic matter is accumulated, microorganisms break down the feedstock by consuming carbon and producing heat. The more easily digestible carbon sources are sugars and starch, derived from food wastes. Cellulose and lignin from biomass are more difficult for microbes to decompose and are left behind as a humus-like, organic, nutrient-rich compost that is used for fertilizer and soil amendment.

As microbial activity increases, more heat is released, and the temperature of the compost pile increases, thus increasing microbial activity even more. Therefore, temperature is used to identify the types of microbes with the highest biological activity. As composting is initiated, mesophilic organisms begin aerobic decomposition. As the temperature increases, thermophilic organisms take over, and the rate of decomposition increases. Eventually, the available biodegradable material is consumed, and the microbial activity decreases, as does the temperature. Microbial activity stops, and the compost reaches ambient temperature.

Mechanical mixing, or turning, of a compost pile maintains a high enough oxygen concentration to support microbial activity. If oxygen is not available, then anaerobic conditions exist, and different biological processes are favored. Anaerobic processes are the biological equivalent of partial oxidation. Waste is converted into biogas, a mixture of equal parts methane and CO_2, and other trace impurities like hydrogen sulfide, nitrogen, and ammonia. Biogas can be combusted in an engine generator to produce electricity or cleaned up and recovered as renewable natural gas (RNG) that can be compressed and stored or injected into a natural gas pipeline.

Anaerobic waste conversion is dominant in landfills as impermeable layers of municipal waste accumulate and are left undisturbed. Also, many landfills are lined and covered to manage leachate runoff and air emissions, respectively. Anaerobic digesters are also closed systems used to convert wet wastes, such as animal manures, food wastes, and biosolids, into biogas. Anaerobic digestion diverts the large volume of food and other organic waste that would otherwise go to a landfill. Economic benefits for waste producers, such as wastewater treatment plants and food processing facilities, are realized as reduced hauling cost and lower landfill tipping fees. An example of a 10-L laboratory scale anaerobic digester used for converting residual carbon in wastewater to biogas is shown in Fig. 12.1.

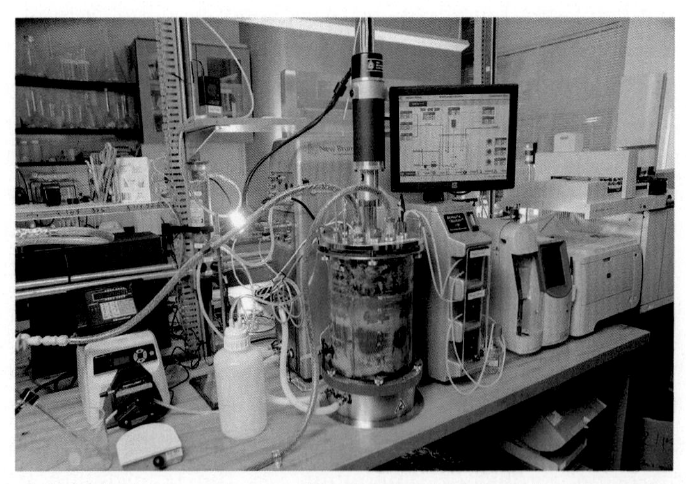

Fig. 12.1 A 10-L laboratory-scale anaerobic digester used to study carbon recovery from wastewater to optimize biogas production. *(Photo by John Theilgard, RTI International.)*

Lower waste disposal cost is not enough to offset the capital investment in new anaerobic digestors. The shift from landfilling to digesting results in value added products such as renewable energy in the form of RNG that can be used to produce biopower or as an alternative transportation fuel (compressed RNG). Recently, RNG has been qualified as a renewable transportation fuel that is eligible for RIN credits, as discussed in Chapter 8, providing additional economic incentive for biogas upgrading to RNG.

Biogas composition is a function of how and from what it is produced. Landfill gas contains 40–60% CH_4, 25–40% CO_2, up to 17% nitrogen, around 2% oxygen, and several hundred ppm of H_2S. Biogas from anaerobic digestion of biosolids and other organic wastes contains higher methane (60–65%), comparable CO_2 (~40%), similar H_2S, but much lower nitrogen (1–8%) and only a trace of oxygen.[8]

Biogas requires upgrading and purification before it can be classified as RNG. Several commercial technologies, such as pressure swing adsorption, physical and chemical absorbents, membrane separation, and cryogenic separation, are available to meet natural gas pipeline standards (90–98% CH_4, 4–10 ppm H_2S, <1% oxygen, and no particulates). Pressure swing adsorption is a common, cost-effective technique to upgrade biogas on a large scale.

Externalities

Environmental regulations, policy, and legislation, and social acceptance play at least as much of a role as technology in waste management. Air pollution from waste incineration and groundwater pollution from landfills were significant drivers for new policies that framed environmental regulations. Incinerators are now required to have state-of-the-art air pollution controls to reduce acid gas, particulates, heavy metals, and dioxins. Sanitary landfills are being constructed with liners to minimize leachate runoff.

Concerns about global climate change caused by greenhouse gas emissions are guiding decisions on present and future waste management. Capturing methane from landfills (landfill gas) is a recognized practice that not only facilitates energy recovery from waste but reduces greenhouse gas emissions. At the very least, greenhouse gas emissions are reduced by a factor of 25 by simply flaring methane to produce CO_2. Methane has 25 times greater greenhouse warming potential than CO_2. Recovering landfill gas has even greater potential for greenhouse gas emissions reduction if it offsets fossil fuel use for energy production.

The fundamental goal of waste management is to protect the environment while preserving economic prosperity to improve quality of life. This goal remains constant even as society evolves. Old problems are solved while new challenges are created or discovered. Waste streams continue to increase in volume and become more complex. Consequently, legislators need to adapt their risk-based decision-making processes to enact comprehensive policies reflecting current societal needs with enough flexibility to anticipate future needs.

Consistent policy creates a stable regulatory environment that reduces the financial risk of capital investment for improving existing infrastructure and developing new technological solutions. It also instills greater public awareness and confidence for social acceptance. The perception of waste as a disposal problem is evolving into waste as a resource. This creates value for waste as a source of energy and justifies infrastructure development to increase profits. Policies and regulations that incentivize waste reduction by reduced consumption, reuse, and recycling could ultimately affect the value of waste for energy production and threaten future investment. Of course, promoting wastefulness to increase resources for waste to energy is not the intended consequence, but it does highlight the fact that the complexity of integrated waste management, if not carefully controlled, can result in conflicting circumstances.

Is a Circular Economy the Solution for Waste Management?

The paradigm of waste to energy grew out of the necessity to manage waste in a linear economy where natural resources are transformed into raw materials that are converted into products. Value is increased along this chain until goods are purchased by the consumer. The owner of the product decides the duration of utility and the fate of the product at end of life. The consumer's decision to reuse, recycle, or dispose of materials is driven by cost.

Enter the circular economy.[9] The goal is to maximize the value at each point during product life. As succinctly stated by Stahel, "The goods of today become the resources of tomorrow at yesterday's prices."[9] By design, the circular economy is based on minimizing input energy and raw materials by extending the useful life of goods and materials while reducing waste. This requires maximizing system efficiency, and therefore cost, at all points in the value chain. Starting at the beginning, goods and products need to be designed for reuse and recycling. This requires efficient, cost-effective infrastructure and technologies to collect materials for reuse and recycling, in addition to developing secondary markets for reused or refurbished products and recycled materials. Overarching public policy and regulations need to be in place to support the transition to a circular economy by encouraging public adoption.

Reusing or repurposing is cheaper and easier than recycling and producing new products if there is established infrastructure to collect, refurbish, and redistribute or resell into secondary markets. For example, glass beverage bottles are returned to the distributor, washed, and refilled for resale. The advantage to the consumer is in paying for only the product, not the product and the container, but the responsibility rests with the consumer to return the container for reuse. Recycling materials generally requires less energy input than recovering raw materials from natural resources. Glass and aluminum containers are two good examples; there are many more. Recovering, remelting, and reforming aluminum containers is less energy intensive than smelting bauxite to produce virgin aluminum for manufacturing new containers.[10] Recycling plastics is a bit more complicated because of the different types of plastic. Effective plastics recycling is heavily dependent on cost-effective presorting.

The circular economy strives to reduce environmental impact by lowering greenhouse gas emissions from energy consumption and reducing waste disposal. This may conflict with technologies that rely on waste as a feedstock to produce energy. Conventional waste management is driven by reducing the cost of disposal, but waste management companies profit from waste disposal, and the cost is passed on to the consumer. The importance of new business models across the entire value chain from product development through waste disposal are needed to redistribute the value of reducing environmental impact. This may require new social paradigms and technical innovations to foster a more economically and environmentally sustainable society.

Close-up: The Circular Economy

The journal *Nature* recently had an entire section devoted to the Circular Economy (Fig. 12.2).[9–11] It is a concept that is gaining currency, largely through the efforts of the Ellen MacArthur Foundation (https://www.ellenmacarthurfoundation.org/publications). In brief, it stands for a society with a minimum of waste. The grand vision is to decouple economic growth from resource consumption. If this seems like a tall order, it is. But that does not mean we ought not to set such an aspirational goal.

Continued

Close-up: The Circular Economy—cont'd

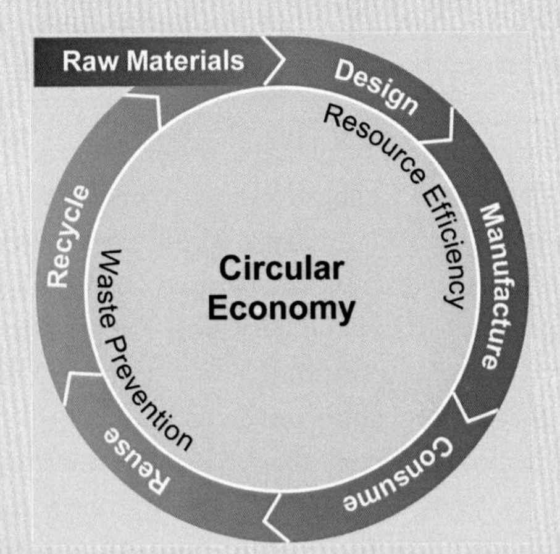

Fig. 12.2 The circular economy.

When society is premised upon the continuum of take-make-use-refuse, then economic growth comes with the use of more resources. It begins with the resource in the ground, be it the metal for objects, gypsum for wall boards, or petroleum for fuel or plastics. Energy, mostly derived from non-renewable fossil fuel, is used to make things. The produced goods are used, sometimes consumed, and most often discarded as waste after the use.

The circular economy seeks to minimize waste at each of the steps. Manufacturing today discards up to 90% of the raw material. Certainly, additive manufacturing (3D printing) will help where it can. But the design step before that is just as important. Most anything can be designed for repair, refurbish, and re-use in general, or not. This underlines a key tenet of the concept. It is very much not only about recycling. Recycling is often very energy inefficient, while it does conserve hard resources such as metals. Besides, recycling requires the step of aggregation, and this is often uneconomic.

The most successful model of recycling is that of lead from automotive batteries. The system made it happen. Consumers were given a discount on the new battery when returning the old one rather than discarding it. In states and provinces that do it, beverage containers (glass and plastic bottles and aluminum cans) have a deposit returned. One could argue whether the 5–25 cents drives behavior or not. But the principle is one worth emulating. A power drill returned for refurbish and subsequent sale ought to have a value set on the return. But it would in the first place have been designed

Close-up: The Circular Economy—cont'd

for economical refurbish. This underlines a key element of the circular economy: emphasis on repair, refurbish, and re-use in place of buying new. Expenditure on resources would be replaced with expenditure on labor. That means jobs.

Plastics ought to be an area of emphasis. The vast bulk of them are derived from natural gas or oil. The United States discards 35 million tons per year and either recycles or combusts only 14.3%. They almost symbolize the use and discard concept and are the epitome for economic growth being proportional to resource consumption. An estimated 100 million tons of plastic waste is floating around in the oceans. Part of the problem is the variety of plastic formulations used and not necessarily sorted. Only two, PET (polyethylene terephthalate) and high-density polyethylene are routinely recycled. The research emphasis has been on recycling, although attempting to reduce the energy to recycle is a key aspect of a developing circular economy. In the United States, only 34% of plastic is recycled. However, in some states there are laws that allow a deposit on beverage containers. At the time of purchase, the consumer pays a 5–10-cent deposit that is returned when the bottle is returned. In states with bottle deposit laws, plastics recycling increases to 70%, and it is almost 90% in Michigan, which has the highest bottle deposit (10 cents).[12]

In the end, it may not only be about technology. Business model change could drive behavior. If we went to more of a lease economy than buy and sell, the leasing entity would be advantaged by the equipment being maintainable and by it simply lasting longer. If they are also the manufacturers, they are positioned to design to achieve that result. Modern sensing and communication techniques would give them early warning for preventive maintenance. Similarly, sharing versus owning could play a part. The Airbnb model could be broadened to short term rentals of underused capital such as chain saws and party coolers. Sophistication in data analytics, credit management, and communications allows this to be done cheaply.

Reasonable atmospheric carbon dioxide reduction targets can only be achieved by simply using less energy for the same utility. The circular economy concept is a model for alleviating future carbon-constrained growth and development.

Vikram Rao

Research Triangle Energy Consortium, Research Triangle Park, NC, United States

References

1. Hoornweg D, Perinaz BT. *What a Waste: A Global Review of Solid Waste Management*. vol. 15.
2. Makarichi L, Jutidamrongphan W, Techato K-a. The evolution of waste-to-energy incineration: a review. *Renew Sust Energ Rev*. 2018;91:812–821.
3. McCarty PL, Bae J, Kim J. Domestic wastewater treatment as a net energy producer–can this be achieved? *Environ Sci Technol*. 2011;45(17):7100–7106.
4. Seiple TE, Coleman AM, Skaggs RL. Municipal wastewater sludge as a sustainable bioresource in the United States. *J Environ Manag*. 2017;197:673–680.
5. Skaggs RL, Coleman AM, Seiple TE, Milbrandt AR. Waste-to-Energy biofuel production potential for selected feedstocks in the conterminous United States. *Renew Sust Energ Rev*. 2018;82:2640–2651.
6. Langholtz MH, Stokes BJ, Eaton LM. *2016 Billion-Ton Report: advancing domestic resources for a thriving bioeconomy*. EERE Publication and Product Library; 2016 DOE/EE-1440; Other: 7439 United States 10.2172/1271651 Other: 7439 EE-LIBRARY English.
7. Arena U. Process and technological aspects of municipal solid waste gasification. A review. *Waste Manag*. 2012;32(4):625–639.
8. Ullah Khan I, Hafiz Dzarfan Othman M, Hashim H, et al. Biogas as a renewable energy fuel—a review of biogas upgrading, utilisation and storage. *Energy Convers Manag*. 2017;150:277–294.
9. Stahel WR. Circular economy. *Nature*. 2016;531(7595):435–438.
10. Geng Y, Sarkis J, Bleischwitz R. Globalize the circular economy. *Nature*. 2019;565 (7738):153–155.
11. Circular economy: getting the circulation going. *Nature*. 2016;531:443–446.
12. Leone G, Borras JMT, Marttin EV. *The New Plastics Economy: Rethinking the Future of Plastics & Catalyzing Action*. Ellen Macarthur Foundation; 2017 December 13, 2017.

Additional Reading

Leone G, Borras JMT, Marttin EV. *The New Plastics Economy: Rethinking the Future of Plastics & Catalyzing Action*. Ellen Macarthur Foundation; 2017 December 13, 2017.

CHAPTER THIRTEEN

Carbon Dioxide Utilization for Building Materials, Fuels, Chemicals, and Polymers

Introduction

Reducing carbon dioxide (CO_2) emissions and the atmospheric CO_2 concentration to levels recommended by climate scientists is a very significant challenge. Carbon capture and utilization (CO_2U) has the potential to be a significant part of the solution. These technologies have the benefit of reducing direct CO_2 emissions, capturing CO_2 in long life forms such as building materials, and displacing the source of fossil fuel carbon emissions if the CO_2 is converted into fuels and chemicals that would normally be produced from a fossil feedstock. Geological sequestration of carbon will always be an added cost and so is not driven by market factors. Capturing CO_2 from the air or from point sources such as power plants and ethanol mills and turning it into useful products generates a market for CO_2, making it possible to monetize the environmental benefits of CO_2U.

Current Global Usage of CO_2

Global consumption of CO_2 was 206 million tons in 2017, of which 29 million tonnes (14%) were utilized in liquid or solid form and the remaining 177 million tonnes in the gaseous form used mostly for enhanced oil recovery and the production of urea (nitrogen fertilizer). Fig. 13.1 gives a breakdown of CO_2 markets with actual figures for 2014 and 2105 and yearly projections through 2025. The global CO_2 market value was $6.0 billion in 2015, with a projected compound annual growth rate of 3.7% for the next 10 years. Nearly all of this growth will be driven by increased use for enhanced oil recovery. Significant growth of CO_2 use in the food and beverage market is not expected because of societal factors such as reduced per capita consumption of carbonated beverages and the increasing desire for

Analytical Methods for Biomass Characterization and Conversion
https://doi.org/10.1016/B978-0-12-815605-6.00013-5

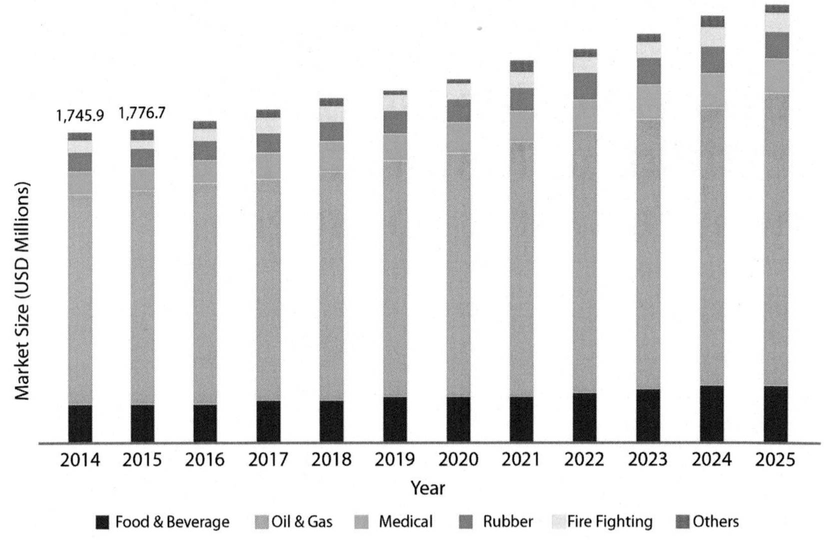

Fig. 13.1 CO_2 markets projected through 2025.

fresh over frozen food, as a significant amount of CO_2 is used for freezing in the food industry. Medical, firefighting, and traditional industrial uses account for only about 0.6% of worldwide CO_2 emissions, but that equates to 37 billion tonnes. This highlights the fact that additional markets need to be developed to increase CO_2U by at least 10-fold to have an impact on global CO_2 emissions reduction.

Potential Future Markets for CO_2

Future markets could fall into two categories: non-reductive, where the CO_2 is used in either gaseous or liquid form, and reductive, where the CO_2 is reduced to form a reactive intermediate such as formate, carbon monoxide (CO), or methanol, that can be used for fuel and chemical production. There are electrochemical, biological, and thermochemical methods for reducing CO_2 to a reactive intermediate. To be a true CO_2U technology, it is critically important that the reductive results in a process with net negative CO_2 emissions.

Non-Reductive Markets

A comprehensive study by the Global CO_2 Initiative at the University of Michigan[1] identified building materials as a growth market. This study highlighted the potential of concrete and carbonate aggregates to sequester a large amount of CO_2 in stable materials for a long time. Of particular interest is the potential of new CO_2 enhanced concrete to sequester even more CO_2 than traditional concrete. For aggregates, the CO_2 is reacted with calcium (Ca) and magnesium (Mg) to make carbonates consisting of calcite ($CaCO_3$), magnesite ($MgCO_3$), or mixtures of the two.

Concrete

In 2018, the concrete industry accounted for about 8% of global carbon emissions.[2] Most of these emissions are associated with the clinker formation process in portland cement manufacturing. Clinker contains various calcium silicates that serve as the binder in cement. It is made by heating a homogeneous mixture of raw materials in a rotary kiln at a sintering temperature of about 1450 °C. The clinkers are then finely ground with added calcium sulfate to form the finished cement powder that is mixed with aggregate (gravel and sand) and water to form concrete. The concrete is poured or cast for such products as sidewalks and concrete buildings and allowed to cure by a series of reactions, including the conversion of calcium hydroxide to calcium carbonate by absorption of CO_2 from the atmosphere.

Carbon dioxide enhanced concrete is relatively new. Controlled amounts of CO_2 are injected directly into the ready-mixed concrete, usually in the concrete truck during the initial batching and mixing. When liquefied CO_2 is injected into wet concrete it chemically reacts with calcium ions released from the cement in the concrete to form solid, nano-sized calcite particles that become permanently bound within the concrete. These stable particles in the concrete matrix can increase the compressive strength of concrete by up to 20% and sequester the CO_2 for the life of the concrete, which can be decades or even centuries. The process also reduces water usage and the overall concrete manufacturing cost. Unfortunately, market adoption has been slow because the concrete industry is reluctant to incorporate new technologies and methods that might reduce product quality.

The University of Michigan analysis showed that the CO_2 enhanced concrete market could increase CO_2U from ~40 million tonnes in 2018 to 600 million in 2030, or to 1.4 billion by 2030 if favorable policies and tax incentives were implemented. This would be a 15- to 35-fold increase.

Aggregates

Aggregates are mined from quarries and used as fill material in concrete, asphalt, and construction. Carbonate aggregates are an alternative that could be made using CO_2 to convert low-value materials such as solid waste containing calcium oxide, slugs from steel mills, and ash from municipal incinerators, into useful products. This would have the dual benefit of sequestering CO_2 and providing a value-added market for waste streams that currently go to landfill. That would need to be balanced against the CO_2 emitted from transporting the solid waste and the CO_2 to the manufacturing site.

Aggregates are fill materials that have low value with minimal performance specifications. The quality of carbonate aggregates would essentially be equal to that of conventional aggregates mined from quarries. Hence the cost of carbonate aggregates needs to be equal to or less than the cost of conventional aggregates to enter the market, or favorable policies encouraging the use of these aggregates would need to be in place.

The University of Michigan analysis concludes that market growth projections for carbonate aggregates are very dependent on favorable policies. Estimates for the potential CO_2 sequestration in carbonate aggregates is 10 million tonnes in 2018 and 300 million in 2030, or up to 3.6 billion with favorable policies and tax incentives. This is a 30- to 360-fold increase in CO_2U potential.

Reductive Markets

CO_2 is the fully oxidized form of carbon, with no heating value and very low reactivity at low temperatures. Therefore, the first step to using it as a feedstock for materials, fuels, chemicals, or polymers is to reduce it to a reactive intermediate. It can be reduced by the Boudouard reaction at a minimum of 700 °C, but there are several ways to do this using renewable electricity at lower temperature. Fig. 13.2 illustrates reduction to a reactive intermediate with low-cost wind and solar power to drive the process without any associated CO_2 emissions. The mismatch in the time of day between peak wind and peak solar generation and electricity demand can also be used as an advantage to design a reduction process that would access the most economical power source and point on the demand curve. The design of the reduction process will also need to accommodate intermittent wind and solar power production.

Direct Electrochemical Reduction

In this method, electrons are directly supplied to an electrochemical cell where CO_2 is reduced at the cathode by protons which are produced by water oxidation at the anode and transported to the cathode via a selective

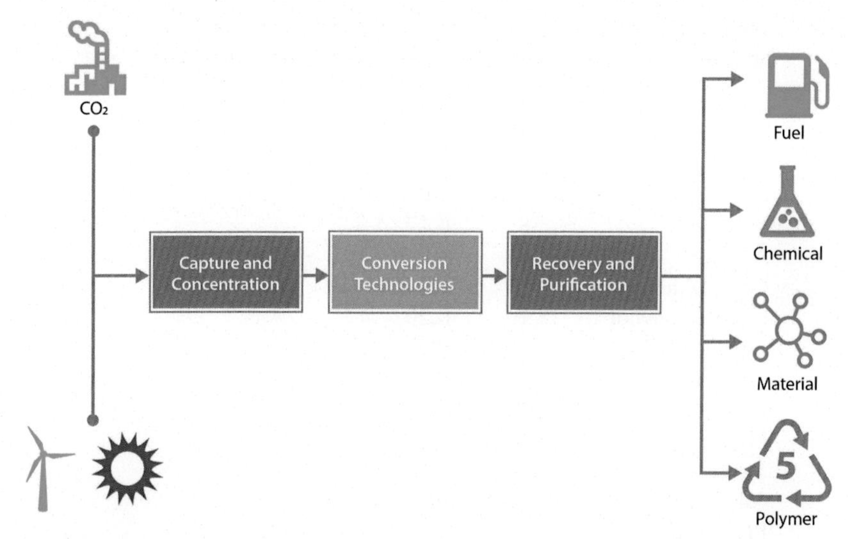

Fig. 13.2 Wind- and solar-driven CO_2U to produce fuels, chemicals, materials, and polymers.

membrane (Fig. 13.3). One attractive feature is that a wide range of reactive intermediates can be generated (see the right side of Fig. 13.3) depending on the number of electrons supplied. The applied cell potential, electrolyte, and catalyst properties can be manipulated to produce a large number of reduced

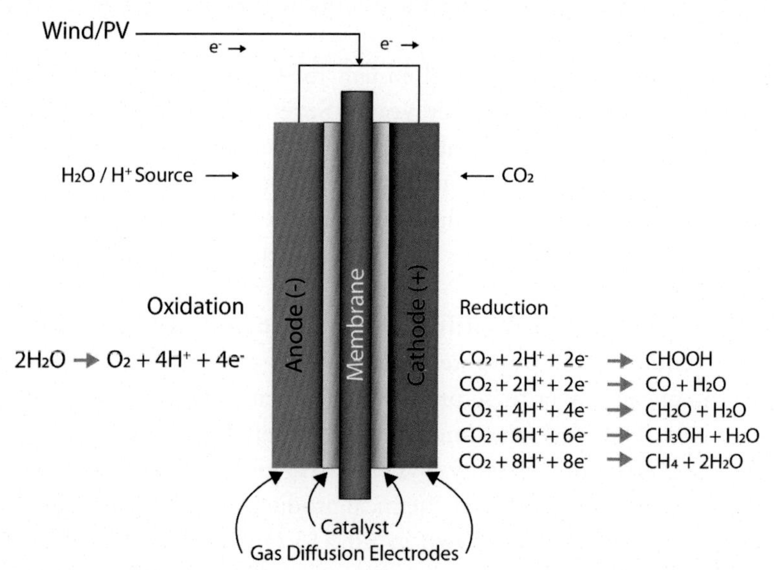

Fig. 13.3 A membrane electrode assembly for common oxidation and reduction reactions.

Table 13.1 Observed Products in the Electroreduction of CO_2.

Species	#e-	Reduction potential vs RHE (V)[a]
Carbon monoxide	2	−0.10
Formate	2	−0.02
Methanol	6	0.03
Glyoxal	6	−0.16
Methane	8	0.17
Acetate	8	−0.26
Glycolaldehyde	8	−0.03
Ethylene glycol	10	0.20
Acetaldehyde	10	0.05
Ethanol	12	0.09
Ethylene	12	0.08
Hydroxyacetone	14	0.46
Acetone	16	−0.14
Allyl alcohol	16	0.11
Propionaldehyde	16	0.14
1-Propanol	18	0.21

[a]Reduction potential provided vs reversible hydrogen electrode (RHE) at pH 6.8.[1]

reactive products containing one, two, and three carbon atoms (C_1, C_2, and C_3 compounds) with a variety of chemical functionalities, as shown in Table 13.1. Another advantage is the relative ease of starting and stopping the process, making it suitable for the intermittent nature of wind and solar energy.

Research on technology development for CO_2 reduction by electrochemical methods dates back 30+ years.[3] Several approaches, notably the reduction of CO_2 to CO, are entering the marketplace. However, the formation of the higher carbon-containing compounds and the higher value compounds still faces technical challenges because of more complex electrochemistry. Following are three of those challenges, with suggestions to overcome them.

- **High overpotential resulting in energy losses**: Many of these processes require a high overpotential. While off-peak, low-cost renewable electricity can power these processes, electricity still represents a significant cost. Therefore, maximizing overall energy efficiency is critical to achieving economic viability. System energetic efficiency is the ratio of energy stored in the reduced chemical product to the total energy input into the system. This is shown in Eq. (13.1), where E_i^0 represents the thermodynamic potential, $\epsilon_{i,\text{Faradaic}}$ represents the faradaic efficiency (FE) of the *i*th species, and η represents overpotential.[4] In this context, faradaic efficiency serves as a proxy for chemical selectivity within the electrochemical cell and represents the ratio of electrons used for the desired products to the total electron input.

$$\epsilon_{energetic} = \sum_i \frac{E_i^o \epsilon_{i,\,Faradaic}}{E_i^o + \eta} \quad (13.1)$$

- **For C_2 and above compounds, forming C—C bonds with high faradaic efficiency**: Electrocatalysts used for reducing CO_2 to C_2 products and above need to balance the binding of CO to the active catalyst site without blocking C—C coupling. Based on research to date,[5] copper is the best monometallic electrocatalyst that is capable of forming C_2 and above species, with a total FE >1%. Significant improvements have been made on this, up to 57–70% cumulative FE, for a class of oxide-derived copper catalysts comprised of continuous metal films containing 10–100 nm crystallites. Other approaches such as varying the size of the metal cationic species used in the solutions have shown improvements in FE to C_{2+} products approaching 70%, so future research to understand and improve the effects of surface morphology on product distributions and to improve catalyst structure should show additional improvements.
- **Electrochemical cell design, scalability, and durability**: Due to its ability to function at mild operating conditions and its ease of operation, the most commonly used electrochemical cell is the aqueous three-electrode system. However, aqueous systems suffer from low solubility of CO_2 and challenges in scalability, which limits commercial viability. The low solubility of CO_2 at atmospheric conditions would greatly limit the current density that could be applied, which would lead to very large electrochemical cells being required at high capital cost. Gas-phase or pressurized designs such as membrane electrode assembly, gas diffusion electrode, or solid oxide electrochemical cell would allow higher current densities to be applied at good energy efficiency, which would allow smaller electrochemical cells and lower capital cost.

Direct Bioelectrochemical Reduction

Whereas electrochemical techniques rely on metal-catalyzed reactions for the CO_2 reduction and product formation, an alternative uses microorganisms. Microbial electrosynthesis (MES) in theory offers advantages over direct electrochemical cells: (1) lower overpotentials required to activate CO_2 reduction; (2) higher selectivity to C_{2+} products, theoretically up to 100%: (3) greater flexibility to desired products enabled by advances in

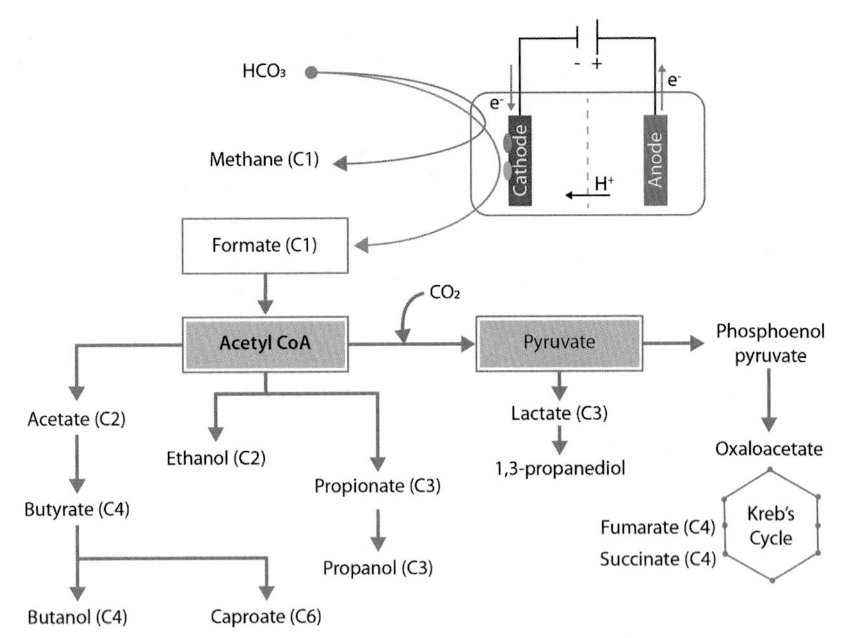

Fig. 13.4 A microbial electrosynthesis cell and accessible synthesis products via the Wood-Ljungdahl pathway.

systems biology to genetically engineer biocatalysts; (4) lower operating cost; and (5) inherent regenerative properties of the biocatalyst. However, a major concern is lower throughput. These benefits and concerns will need to be balanced.

Fig. 13.4 is a conceptual representation of an MES cell and available products using the biocatalyst on the cathode. In the process known as extracellular electron transfer, specialized electroreactive bacteria—"electrotrophs"—directly use electrons to reduce CO_2 to products. This process can be direct or indirect. In direct transfer, the biocatalyst is in direct contact with the electrode surface whereby the uptake of electrons proceeds through nanowires or exposed surface cytochromes which convey them inside the cell body. An alternative mechanism is a multi-step process know as mediated electron transfer, in which separate reversible redox mediators or "electron shuttles" (e.g., H_2, formate, Fe(II), and NH_3) transfer electrons between the biocatalyst cells and the electrodes.[6]

For MES design, it is crucial to understand what electron transfer method is occurring, because pathways are very different. In direct electron transfer, MES productivity will be limited by the surface area of the electrode and the effective coverage of the biofilm. Productivity in a MES system is also controlled by the concentration of biocatalyst and mediators in the solution, solubility of CO_2 and other reactants, and current flow.

Electrochemical reduction approaches have received considerable interest and development over the past 30+ years, but there has been relatively little work on MES approaches. However, MES has considerable promise because of technical advantages and recent advances in systems biology. Following is an overview of the challenges for MES.

- **Better understanding of electron transfer mechanisms, leading to better control**: The flow of electrons within electrotrophs is not well understood for either direct or mediated transfer. Much of the current knowledge is based on a few studies for a few organisms that act as electron-donating species. Previous studies have shown that extracellular electrotransfer mechanisms vary between bacterial species and are different for electron donating vs electron accepting. The feasibility of direct transfer and the role of H_2 formation in electron transfer is still debated. Without greater scientific understanding of electron transfer mechanisms, the utility of systems biology techniques for improving the biocatalyst in MES will not be effective.
- **Slow reaction rates**: Rates are at least two orders of magnitude slower than in electrochemical cells. Several approaches to increase them are being pursued, such as developing biocatalyst compatible cathodes, which maximize surface area for biofilm coverage and are also efficient in delivering electrons and H_2. Reaction rates in MES systems are also a function of the applied current density (CD). Unfortunately, the CD that can be applied is limited by the stability of the biofilm. Developing biofilms that are stable at high applied CD would be very advantageous.
- **Product separation and toxicity**: A major advantage of MES is high selectivity to desired products. Therefore, a pure and concentrated product stream must be achieved. Above a certain concentration, a biocatalyst will be inhibited by the product. Implementing in situ separations can mitigate product inhibition, but it must be done in a cost-effective manner. Systems biology techniques can be applied to limit the inhibition of the biocatalyst to the product.

Indirect Bioelectrochemical Reduction

This approach (Fig. 13.5) involves the electrochemical production of an electron donating mediator species such as H_2 that is co-fed with CO_2 to the biocatalyst in a downstream bioreactor. H_2 acts as a reducing agent,

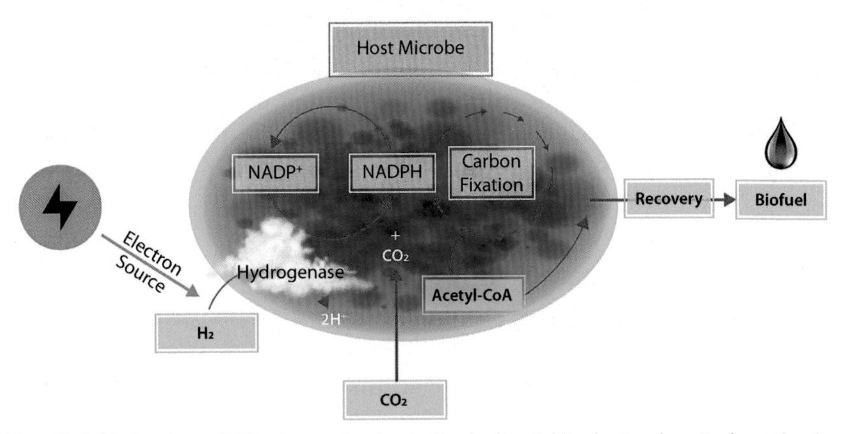

Fig. 13.5 Reduction of CO_2 to products via the indirect bioelectrochemical method.

in contrast to the direct reduction discussed in the previous section, where the electrons were directly used by the biocatalyst. The advantage of this approach is that separating the faster process step (producing renewable H_2 from water electrolysis) from the slower bioreactor step makes it easier to integrate intermittent low-cost renewable electricity. However, the electrolyser adds capital and operating costs.

Because the techniques for indirect biochemical reduction are more developed and better understood, the technical challenges for commercial implementation are lower compared to direct techniques and center mostly around the low solubility of CO_2 and H_2 and safety concerns about feeding H_2 into anoxic conditions. The same product separation and toxicity challenges exist for indirect as for direct bioelectrochemical reduction.

Indirect Thermochemical Reduction

Thermochemical approaches such as reverse-water gas shift (rWGS) (Eq. 13.2), steam methane reforming (SMR) (Eq. 13.3), and biomass (BG) or waste carbon gasification (Eq. 13.4) are known commercial processes for reducing CO_2 to CO, producing H_2 or syngas (CO and H_2).

$$\text{rWGS}\ \ CO_2 + H_2 \leftrightarrow CO + H_2O \tag{13.2}$$

$$\text{SMR}\ \ CH_4 + H_2O \leftrightarrow CO + 3H_2 \tag{13.3}$$

$$\text{BG}\ \ C + H_2O \leftrightarrow CO + H_2 \tag{13.4}$$

Once CO and H_2 are produced, several proven techniques are available for converting them to C_{2+} value added products, such as Fischer-Tropsch synthesis (FTS), methanol to gasoline, methanol to olefins, and other less well-known processes. Although it is possible to use zero-carbon-emission renewable electricity and CO_2 as the feedstock, so that they qualify as CO_2U, it is not clear how this could be accomplished at commercial scale and be economically competitive without significant policy incentives. Existing commercial processes use mostly fossil fuel feedstocks and energy sources, and extensive technical modifications would be required to use renewable electricity sources and CO_2 to make this a viable CO_2U approach.

Markets for Reduced CO_2 Products

The major market categories are chemicals, fuels (methane and liquid fuels), and polymers. Following is a description of each market and its potential.

Chemical Intermediates

A multitude of chemicals can be produced from CO_2. Market potential can be predicted by looking at the cost of production from CO_2 compared to the current market price and then evaluating market size. Table 13.2 shows the cost of production for various C_1 and C_2 products. This analysis was based on an electricity cost of $0.03/kWh for off-peak demand for wind and solar. Some very interesting observations can be drawn. The first three products,

Table 13.2 Production Cost vs Market Price for Various C_1 and C_2 Chemical Products.

	No. of electrons per mole	Target faradaic efficiency (%)	Target cell voltage (V)	Target current density (mA/cm^2)	Production cost ($/kg) at target value	Market price ($/kg)	Required CO_2 credits to reach parity ($/tonne)
CO	2	98	1.93	1500	0.22	0.23	$0
Formic acid	2	98	2.08	1500	0.25	0.63	$0
Oxalic acid	2	95	2.30	1500	0.44	1.73	$0
Ethylene	12	95	1.82	1500	1.15	0.57	$150
Ethanol	12	95	1.74	1500	0.70	0.51	$70
Methane	8	95	1.66	1500	1.24	0.22	$340
Methanol	6	95	1.80	1500	0.52	0.40	$65

from C_1 and C_2, are economically viable without any policy incentives, such as a carbon credit. Products that require more than 2 electrons per mole for conversion are not competitive under the aggressive but achievable conditions shown. The far-right column is the carbon credit required to achieve market price parity. The value of this credit ranges from $65 to $340 per tonne of CO_2.

CO, formic acid, and oxalic acid all show good potential for economic competitiveness. CO or syngas ($CO + H_2$) is very versatile as a chemical intermediate and can be used to make many finished chemicals or liquid fuels by, for example, FTS. Current market size is relatively small but is growing at a compound rate of 8%. With policies in place to encourage CO_2U, syngas could see rapid growth, given its versatility to produce a wide range of fuels and chemicals and its amenability as a relatively low-cost add-on to existing chemical plants and refineries.

Formic acid, a C_1 carboxylic acid, looks very attractive, with a targeted production cost less than half of the current market process. It is a very versatile intermediate because of its strong acidity. It is used as a preservative and antibacterial agent in silage for livestock, tanning of leather, dyeing in textiles, and cleaning products. Its many uses commands a high price, $800 to about $1250 per tonne. It is commercially produced mostly from methyl formate and formamide. Although the markets are attractive, they are relatively small, and none of the existing uses has the potential to grow at the scale necessary for significant CO_2U. It has been proposed as a fuel cell, but much development would be necessary to make this application viable.

Oxalic acid, a C_2 dicarboxylic acid, also looks very attractive, with production costs less than one-third of current market prices. It is made mostly by the oxidation of carbohydrates or glucose and used predominantly in cleaning and bleaching. However, as with formic acid, none of the current uses has potential for CO_2U at scale.

Ethylene is a versatile intermediate with a very large market, notably for making polyethylene. But the conversion cost is over double current prices.

Although there might be other opportunities, the best seem to be in products that require the least number of electrons for synthesis, ideally for the synthesis of chemicals with the highest market price. Given the orders of magnitude mismatch between the chemicals markets and anthropogenic CO_2 emissions, none of these markets will scale at the level necessary to have a significant impact for CO_2U. However, because syngas, formic acid, and oxalic acid show the potential to be market competitive without the need for policy incentives, they are good opportunities for initial deployment of CO_2U technologies.

The University of Michigan study shows the current market size for chemical intermediates produced from CO_2 reduction to be 100,000 tonnes, with a growth potential to 5 million tonnes without policy incentives and 50 million with incentives, by 2030. This is significant growth, but it would have relatively little impact on global CO_2 emissions. The main opportunity for chemical intermediates appears to be the initial one for the shakedown and demonstration of CO_2 reduction technologies.

Methane

Methane (natural gas) from CO_2 has the potential for CO_2U at global scale. Unfortunately, the production cost is over five times the current market price. To reach price parity would require an incentive of $340 per tonne to cover reduction of the 6 electrons per mole. Anaerobic digestion from biomass is more competitive for producing low carbon methane. In short, reducing CO_2 to methane without either a policy incentive or a technological breakthrough faces a significant economic challenge for price parity with natural gas. Currently some markets will pay a significant premium, for instance Southern California, for renewable methane, and this could provide niche opportunities.

Liquid Fuels

The transportation sector is the largest emitter of CO_2 globally and is powered predominantly by gasoline, diesel, and jet fuel derived from crude oil. Use of low carbon biofuels has been growing rapidly but still accounts for $<10\%$ of fuel use globally. Any significant switch from crude oil would have an impact on CO_2U.

Table 13.2 shows particulars for producing methanol and ethanol by CO_2 reduction. Costs are only about 20% above target values, so relatively small incentives would be required. Biological conversion of CO_2-sourced CO to ethanol has been demonstrated by several technology developers and commercial companies (e.g., LanzaTech). Currently methanol production is mostly from syngas, and ethanol production is from corn in the United States or sugar cane in Brazil. Both can be made from lignocellulosic feedstocks at significant carbon reduction compared to their petroleum counterparts, so significant additional potential for CO_2U is low. From a CO_2U perspective, it would be better to focus on gasoline, diesel, and jet fuel made directly from CO_2 reduction.

FTS is a proven commercial technology for making gasoline, diesel, and jet fuel from syngas.[7] There are many commercial applications of FTS for converting natural gas and coal into transportation fuel and chemicals, and it has been proposed for low carbon biofuels. However, FTS involves many process steps and is therefore capital intensive, requiring a large economy of scale. Unfortunately, the scale of biomass availability is five times or more smaller than that necessary for economical application of FTS.

Table 13.2 shows good economic potential for production of CO from the reduction of CO_2. FTS could be viable for making liquid fuels from CO_2 if it were matched with a renewable low-cost, low-carbon source of hydrogen. The volume of CO_2 produced at power plants and refineries, and even potentially by direct air distillation, is consistent with the volumes necessary for FTS viability. At this scale, liquid fuels from FTS would have a significant impact on CO_2U. The major challenge will be producing hydrogen sustainably in a low carbon manner without increasing cost. Most hydrogen is made by steam methane reforming of natural gas, so CO_2 emissions are significant. Making hydrogen from water electrolysis powered by wind and solar electricity is an active area of research, while photobiological and photoelectrochemical water splitting are longer-term options.

There are several other technology options for producing transportation fuels from syngas, such as methanol to gasoline. Another approach that has received a good deal of research interest and initial demonstration is the conversion of ethanol to jet fuel.[8] As noted above, there are many ways to produce ethanol from CO_2 reduction, making this another CO_2U option for producing transportation fuel.

Because of the remaining technical challenges and lack of policy incentives, only about 10 million tonnes of liquid fuels are currently produced from CO_2 reduction. Growth in the chemical and polymer markets is affected by the possibility of favorable policies being enacted, and this is the predominant factor in projecting the future growth of the liquid fuels market from CO_2. Without favorable policies, projected growth according to the University of Michigan study is 70 million tonnes by 2030, a relatively modest sevenfold increase. With favorable polices the growth projection would be to 2.1 billion tonnes, or 210-fold.

Polymers

Although there are many possibilities for innovative approaches to produce existing or new polymers with better properties from CO_2, most polymer markets are small compared to building materials and liquid fuels, so the potential for CO_2U at the global scale is insignificant.

Total CO_2U Potential

Adding up all of the potential CO_2U growth for the non-reductive (aggregates and concrete) and reductive (chemical intermediates, liquid fuels, and polymers) markets yields an increase of 210 tonnes in 2017 and 1 billion in 2030 without policy incentives, 7 billion in 2030 with incentives. The 7 billion is 19% of current global CO_2 emissions of 37 billion, so a significant potential exists.

Analysis Needs

To determine the concentration and purity of CO_2 and the chemical structure of intermediate and final products will require gas chromatography (GC), mass spectrometry (MS), and GC-MS. GC is typically used for separating and analyzing compounds that can be vaporized without decomposition. It can be used to determine the purity of a substance, in this case CO_2, and the levels of other compounds present. MS ionizes chemical species and sorts the ions into a spectrum based on their mass-to-charge ratio. These spectra are then used to elucidate the chemical structures of molecules and compounds. GC-MS is necessary to identify trace amounts of certain compounds such as contaminants that hinder the CO_2 reduction process to desired products.

Some Useful Publications

Given the concerns associated with CO_2 emissions and the impact on global climate change, there has been a dramatic increase in interest in developing and deploying CO_2U technologies. Fig. 13.6 shows the increase in publications on catalysis development for CO_2 reduction from 2005 to 2015. There is a similar trend for other topics related to CO_2U. There are several good high-level overviews[9–11] and specific ones for building materials[12] and chemical intermediates and fuels.[13]

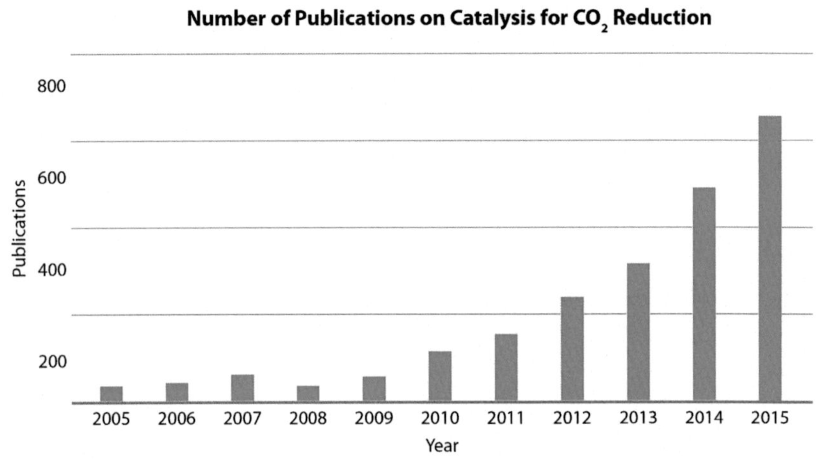

Fig. 13.6 Increase in publications on CO_2 reduction catalysts from 2005 to 2015.

References

1. *Global Roadmap for Implementing CO_2 Utilization.* Global CO_2 Initiative at the University of Michigan; November 2016. http://hdl.handle.net/2027.42/150624.
2. Rodgers L. The massive CO_2 emitter you may not know about. *BBC News.* 2018; 2018-12-17.
3. Hori Y, Kikuchi K, Suzuki S. Production of CO and CH_4 in electrochemical reduction of CO_2 at metal electrodes in aqueous hydrogencarbonate solution. *Chem Lett.* 1985;14(11):1695–1698.
4. Martín AJ, Larrazábal GO, Pérez-Ramírez J. Towards sustainable fuels and chemicals through the electrochemical reduction of CO_2: lessons from water electrolysis. *Green Chem.* 2015;17(12):5114–5130.
5. Clark EL, Hahn C, Jaramillo TF, Bell AT. Electrochemical CO_2 reduction over compressively strained CuAg surface alloys with enhanced multi-carbon oxygenate selectivity. *J Am Chem Soc.* 2017;139(44):15848–15857.
6. Harnisch F, Rosa LF, Kracke F, Virdis B, Krömer JO. Electrifying white biotechnology: engineering and economic potential of electricity-driven bio-production. *ChemSusChem.* 2015;8(5):758–766.
7. Van Der Laan GP, Beenackers A. Kinetics and selectivity of the Fischer–Tropsch synthesis: a literature review. *Catal Rev.* 1999;41(3–4):255–318.
8. Wang W-C, Tao L. Bio-jet fuel conversion technologies. *Renew Sust Energ Rev.* 2016;53:801–822.
9. Yu KMK, Curcic I, Gabriel J, Tsang SCE. Recent advances in CO_2 capture and utilization. *ChemSusChem.* 2008;1(11):893–899.
10. Huang C-H, Tan C-S. A review: CO_2 utilization. *Aerosol Air Qual Res.* 2014;14(2):480–499.
11. Aresta M, Dibenedetto A, Angelini A. The changing paradigm in CO_2 utilization. *J CO_2 Util.* 2013;3:65–73.
12. Jang JG, Kim G, Kim H, Lee H-K. Review on recent advances in CO_2 utilization and sequestration technologies in cement-based materials. *Constr Build Mater.* 2016;127:762–773.
13. Hu B, Guild C, Suib SL. Thermal, electrochemical, and photochemical conversion of CO_2 to fuels and value-added products. *J CO_2 Util.* 2013;1:18–27.

CHAPTER FOURTEEN

Remaining Challenges and Final Thoughts

Introduction

The previous chapters present a broad summary of the many facets of biofuels production. Policies were established in the 2000s (the Energy Policy Act of 2005 and the Energy Independence and Security Act of 2007) to encourage investment and growth in developing a biofuels industry to mitigate global climate change, increase energy security by providing an alternative to fossil fuels, and serve as economic stimulus for rural agricultural communities. The foundation of the industry is what the community refers to as first-generation biofuels: ethanol from corn starch or molasses and biodiesel from vegetable oils. The corn ethanol industry in the United States produced 15 billion gallons in 2017. Fig. 14.1 shows the growth in ethanol production from 1998 through 2017.[1] The United States is the global leader for ethanol production, primarily from corn, followed by Brazil, which uses sugarcane. In 2015, global fuel ethanol production was 25.68 billion gallons: 57.7% by the United States and 27.6% by Brazil.

Biodiesel production in the United States shows a similar trend, but the volume is about one-tenth that of ethanol. The dip in 2010 was caused by increasing uncertainty about the fate of the $1 per gallon biodiesel tax credit that was set to expire in 2012. The Unites States is the global leader, at 18%, followed by Brazil (13%), Germany (13%), Indonesia (12%), and Argentina (12%), which accounts for about half of global production.[2] The relative production volumes are consistent with global transportation fuel usage. Gasoline is predominantly used for light-duty transport in spark ignition (SI) engines and is the most widely used transportation fuel in the world. Diesel is used in compression ignition (CI) engines, which have higher torque and hence load carrying capacity and are appropriate for goods transport in medium- and heavy-duty vehicles. Although CI engines are used in some light-duty vehicles, the future is uncertain predominately because of

Analytical Methods for Biomass Characterization and Conversion
https://doi.org/10.1016/B978-0-12-815605-6.00014-7

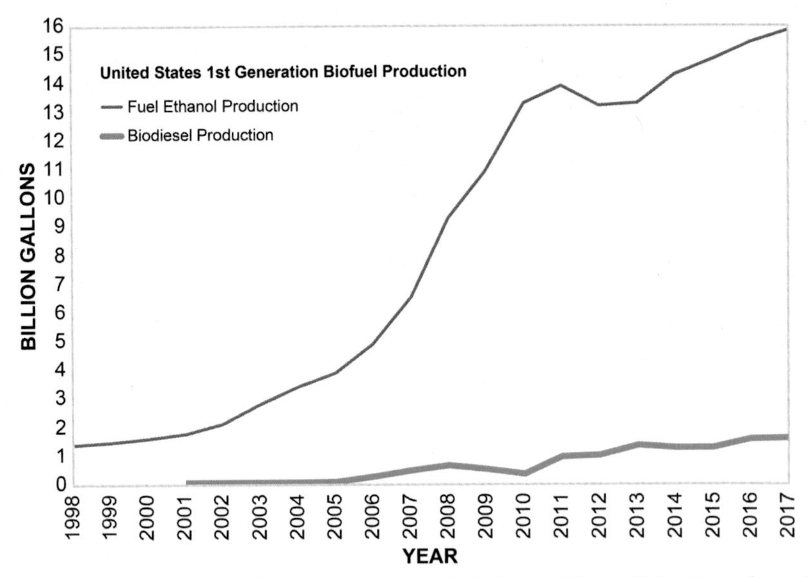

Fig. 14.1 US production of first-generation biofuels from 1998 to 2017. *(Data from the U.S. Energy Information Administration July 2018 Monthly Energy Review.)*

emission concerns. Gasoline use is projected to increase in the near term and then decline steadily due to factors such as decreased vehicle miles traveled and the expected increase in market penetration of electric vehicles. Diesel fuel is projected to grow as a function of world GDP, since the path to electrification of medium- and heavy-duty vehicles should be longer-term than that of light duty vehicles.

Shortages of petroleum in the 2004–07 period caused petroleum prices to rise precipitously, which increased food prices, as fuel is a major cost component of production and transport to market. Biofuel production from mainstay crops—corn, soybeans, palm oil, and sugar cane—was also blamed, spurring the debate of the impact of biofuel production on food prices. The perceived competition for cropland added additional impetus for developing second-generation biofuels based on lignocellulosic feedstocks: woody biomass, agricultural residues like corn stover and wheat straw, municipal solid waste, and construction and demolition waste. Chapter 3 summarized the second-generation technologies to produce lignocellulosic ethanol and advanced biofuels as "drop-in" hydrocarbon replacements. The analytical methods implemented to understand, characterize, and optimize these options were discussed in Chapters 4–6.

Since 2000, there has been significant investment in the research, development, and deployment of second-generation and advanced technologies.

The American Recovery and Reinvestment Act (ARRA) of 2009 provided an unprecedented influx of federal funding (nearly $70 billion[3]) to increase alternative energy production and improve energy efficiency. A significant portion went toward weatherizing homes, businesses, and federal facilities, and extending and increasing energy tax credits. The Recovery Act also established a launching point for biofuels technology by increased R&D funding and support for technology deployment through government sponsored loan guarantee programs. The impact of the funding diminished as projects and programs were sunsetted by 2014 and renewable energy-related R&D returned to pre-ARRA levels.

The commercial demonstration and R&D landscape have changed significantly since 2000. Technical developments in all aspects of advanced biofuel processes are well documented in the scientific literature, as referenced in previous chapters. The focus has also expanded beyond biomass conversion to the biofuels value chain. There is now more emphasis on understanding biomass resource availability, feedstock logistics, environmental sustainability, biofuel specifications for end use, and market acceptance, and how these affect technology development and deployment.

Remaining Challenges

Despite substantial federal and private equity investments in R&D for second-generation technologies, significant technical advances, cost reductions, and several commercial demonstrations, the future of lignocellulosic biorefineries remains uncertain due to technology and market challenges. This section summarizes the state of the art and some of the remaining issues to consider as the biofuels industry continues to mature and evolve.

All Biomass Is Not Created Equal

Determining the physical characteristics and chemical composition of feedstocks were active areas of research in the agricultural and forest products industries long before the advent of the biofuels industry. Cultivar development for maximizing agricultural productivity has been applied to biomass grown specifically for energy-related purposes. Desired properties of selected biomass are dictated by the end use. For heat and power, the most desired feature is energy density; moisture and ash content should be minimized.

Desired features are dictated by the conversion technology. For cellulosic ethanol produced by biochemical conversion, a higher percentage of carbohydrates (cellulose and hemicellulose) compared to lignin is desired because the sugar yield for fermentation is higher for low lignin feedstocks. Mineral matter should be minimized in biochemical conversion because it exacerbates liquid-solid separation, and extractives produce inhibitory compounds that are toxic to fermentation organisms. In thermochemical conversion, higher carbon content and lower oxygen content is desired, so feedstocks with high lignin content are used. Biomass ash content can be an issue for solids handling, and high moisture feedstocks reduce the thermal efficiency of a thermochemical process. Alkali metals and alkaline earth metals are catalyst poisons in the conversion process and downstream upgrading processes. High sulfur and nitrogen content can produce sulfur and nitrogen gases that are catalyst poisons in thermochemical conversion.

Therefore, in general, lower ash content is always desirable, because ash is an inert solid that lowers product yield (on a weight basis), requires disposal, and contributes to catalyst poisoning. Moisture content should be minimized for thermochemical conversion to maximize thermal efficiency. And carbohydrate content should be maximized for biochemical conversion to maximize sugar yield, and ash minimized to reduce impurities.

The variety of analytical methods developed for measuring the physical and chemical characteristics of biomass were described in Chapter 2. The key challenges associated with biomass characterization are sample homogeneity and sample-to-sample variation. There is always uncertainty in the measurements; that can be assessed by using replicate samples and standard reference materials. However, sample-to-sample variability must be considered carefully. For small-scale laboratory experiments, it may be feasible to collect and store a single biomass sample that can be carefully analyzed and used for many years. Larger-scale studies, such as pilot demonstrations, require larger quantities of feedstock (on the order of 1–100 kg/h), so determining the composition of a 1-ton batch may require the collection and preparation of a statistically relevant number of samples to arrive at an average composition with an acceptable range of variance in the measured components. The extent of the variance can be affected by any number of unintended factors.

As an example, if 100 corn stover bales are collected in a single day from a single field that was growing a single variety of corn, one might expect the chemical composition to be very uniform. Now the bales are wrapped in plastic and stored under cover outside a given biorefinery that consumes 5 bales per day. Those bales are processed through a grinder and dried before storage in a feeder bin. The compositional variability of this material will likely be low, assuming no tramp material was trapped inside a bale that inadvertently made it to the process. The packaging and shelter for the bales minimizes compositional changes associated with climatic variations in temperature, wind, and rainfall. However, the longer the bales are stored, even covered, the greater the chance that degradation from microbial activity will induce changes. Fortunately, larger-scale processes typically operate continuously for long periods of time, so subtle changes in feedstock composition are normalized over the time scale of operation. Sample variations can be more disruptive for small-scale laboratory experiments where milligram size samples are converted in seconds, so minimal or no time averaging occurs.

Process performance is usually insensitive to subtle changes in feedstock composition, but specific processes may require specific feedstocks with a narrow range of variability. On the other hand, the cost structure of a mixed biomass resource, such as forest thinnings or mixed hardwoods from logging, may be so attractive that larger variation can be tolerated. Therefore, understanding the relative compositions of different biomass materials is important for advanced technology development. The variance between different feedstocks (wood vs corn stover, hardwood vs softwood, wheat straw vs orchard prunings, etc.) can be quite large. The scientific literature is rich in studies comparing the compositions of different feedstocks, and online databases are available with compilations of the results from many studies and sources.

Understanding how feedstocks behave in each conversion technology has been an active area of research. It may be likely that intermediate or product yields are maximized with a specific feedstock, or the chemical composition of intermediates is more favorable for upgrading into products with one feedstock over another. The caution here is to avoid studying a process with a given feedstock in the laboratory without carefully considering how the results translate to larger-scale systems. On the one hand, a robust scientific understanding of how a feedstock converts to intermediates and products in controlled laboratory experiments is critical for process development based on fundamental physical and chemical principles. However, if this fundamental understanding is to advance the state of biofuels technology, then extrapolating the principles to commercially relevant systems must be considered.

A sensitivity analysis should be performed to highlight the impact of feedstock composition on process performance, and ultimately process economics. Robust technologies that are insensitive to biomass composition may be limited by product yield and quality; selective technologies may require tighter tolerance on composition. Resource availability is an important factor, because a process with restrictions on the types of feedstocks it can tolerate will be limited. In the end, the market will bear the burden, and the value of biomass feedstocks will be determined by their end use.

Biomass Is Not Fossil Fuel

There is a natural tendency to draw parallels between biomass and fossil fuel (coal and crude oil) because of the end-use markets—coal and biomass in electric power generation, crude oil and biomass in transportation. Simply stated, fossil fuels can be considered old biomass that has been pretreated at elevated temperature and pressure over millions of years below the earth's surface, with the primary chemical difference between old and new biomass being oxygen content. Crude oil contains virtually no chemically bound oxygen (<1 wt%), and the oxygen content of coal is low compared to biomass and depends on the rank (age) of the coal. Lignite can contain as much as 30 wt% oxygen, while anthracite can have <5 wt%.

Analytical methods developed for the physical and chemical characterization of coal and petroleum have been adapted for biomass and its conversion intermediates and products. Proximate (ASTM D5142) and ultimate (ASTM D5373) analyses are standard methods for determining the physical (moisture, volatile matter, fixed carbon, and ash) and chemical (carbon, hydrogen, oxygen, nitrogen, sulfur, and ash) characteristics of feedstocks and solid fuels, respectively. Biomass ash can be more volatile and less refractory than coal ash, so the standard methods were modified by using a lower temperature (600 °C) for biomass. Certain biomass materials contain high levels of alkali metals (sodium and potassium) that produce low melting point eutectic mixtures and alkali chlorides, such as potassium chloride; these also have fairly low melting points (KCl 770 °C, NaCl 801 °C). Therefore, certain biomass materials have ash with high vapor pressure at combustion temperatures (~1000 °C), leading to an under-representation of ash content if lower temperatures are not used.

Direct biomass liquefaction intermediates (bio-oil and bio-crude) are often analyzed by standard methods developed for crude oil and petroleum distillates (Chapter 5). However, the only characteristics shared by bio-oils and petroleum are that both are liquids and contain hydrocarbons. Beyond that, there are few if any physical and chemical similarities. Bio-oils tend to be very acidic and thermally unstable because they contain highly reactive oxygenated components. Acidity is measured by pH and total acid number

(TAN). Crude oils have TANs of about 1 and are used to relate the naphthenic acid content to potential corrosion in petroleum refineries. Bio-oils can have TANs of 100 or higher, representing the small molecule organic acid content. This has led to modifications of the TAN methods to reflect more clearly the carboxylic acid content of bio-oils (CAN: carboxylic acid number). Regardless of these modifications, the acidity of bio-derived liquids is one to two orders of magnitude higher than that of crude oils, and the acidic components are chemically very different.

Another laboratory test, Conradson carbon residue (CCR; ASTM D189-06), is a measure of the coke-forming propensity of oils. For petroleum liquids, it gives an estimate of the sooting propensity of diesel fuel, the long-term performance of lubricating oils, and coke formation of a feedstock in a fluid catalytic cracker. The relative thermal instability of bio-oils can be assessed by this technique; however, the chemistry of residual formation from bio-oils is quite different from that of petroleum liquids.

TAN and CCR methods are examples of analyses that can be done on both petroleum and bio-derived liquids. Correlations between petroleum refining process performance and feedstock composition have evolved and improved after years of method development for determining the physical and chemical properties of petroleum liquids. These correlations have not yet been developed for bio-derived liquids. Therefore, discretion is required in interpretation of the results when standard characterization methods developed for petroleum products are applied to bio-oils. The question is not whether the methods can be applied but whether the results are meaningful.

One Person's Trash Is Another's Feedstock

Feedstock resource availability and cost are important considerations for developing biofuels technologies. Unfortunately, the cost of the feedstock is usually directly related to the quality. High quality is defined as a homogeneous composition, low moisture, low ash, and low impurities that can act as catalyst poisons or produce harmful environmental emissions. Quality has a direct impact on the technical risk for developing conversion processes, as lower quality feedstocks can lead to poorer steady-state performance (lower yield) and a greater frequency of unscheduled shutdowns for maintenance. Quality also affects economic risk. High quality feedstocks increase operating costs, and competition for these resources often requires long term supply contracts that can be beneficial if a favorable price is locked in or detrimental if unforeseen changes in product market conditions reduce fuel or chemical prices.

Feedstock cost includes raw material cost and logistics. A feedstock is only useful if it is delivered to the biorefinery, prepared for processing, and stored to maintain an inventory for continuous operation. While considered low quality feedstocks, industrial and municipal wastes are advantaged from both raw material and logistics costs. Any fraction of a waste stream that is not recycled or reused must be disposed of at a cost to the producer; therefore, the raw material cost could be negative, generating an additional revenue stream for the advanced biofuels process developer. Logistics costs can also be greatly reduced or eliminated, because delivering the material to a processing plant or landfill is the responsibility of the waste producer.

The environmental benefit of using waste is another driver. Organic wastes not recycled or reused are a significant carbon resource for energy production that would otherwise be landfilled and potentially emit greenhouse gas and pollutant runoff if not properly managed. Both economic and environmental benefits can be realized by co-locating biomass conversion facilities with waste production and handling facilities, assuming that the process economics are favorable at the proposed scale. Under the right circumstances, there is potential to transform waste liabilities into revenue streams with net positive environmental benefits.

Table 14.1 gives examples of waste feedstocks. The Billion Ton Update estimates that there are up to 80 million dry tons of secondary agricultural residues and wastes potentially available for energy conversion.[4] This estimate does not include residues from food crops such as corn, wheat, and barley.

Table 14.1 Agricultural Residues and Waste for Energy Production.

Secondary cropland residues	*Waste fats, oils, and grease*
Sugarcane residues	Used cooking oils
Cotton gin trash	Yellow grease
Soybean hulls	Brown grease
Rice hulls	Poultry fats
Orchard prunings	Livestock fats
Animal manures	*Municipal wastes*
Dairy manure	Municipal solid waste (MSW)
Swine manure	Wastewater treatment sludges
Poultry litter	Food wastes (residential, commercial, institutional, industrial)

The compositions of end use waste streams like those in Table 14.1 are spatially and temporally variable, requiring additional robustness for the distributed waste-to-energy conversion processes discussed in Chapter 12. Managing the compatibility of the feedstocks for selected processes becomes a significant challenge but is one of the central themes of the developing concept of the circular economy.[5] The conventional linear economy relies on raw materials to make things that are used until the end of their lives and then disposed of. The goal of the circular economy is to maximize the use of raw materials by recovering and reusing products and materials after each end-of-life cycle. Within the circular economy, raw material inputs could be reduced, and waste streams could be minimized or avoided. Ideally, goods and services are produced with the end in mind. Manufacturing processes could be tailored to produce materials that are designed to be reused in much the same way that energy crops are being developed for biofuels. Therefore, agricultural and industrial processes could be purposely developed to provide feedstocks for bioenergy production instead of generating waste.

To Integrate or Not to Integrate, That Is the Question

Biofuels, first-generation or advanced, are considered alternative fuels for a reason. They are not the primary transportation fuel but are a minor component (approximately 10%) of the total transportation fuel market. The infrastructure for production and distribution of transportation fuels was built around petroleum, and alternatives to petroleum must be integrated into this system to be accepted by the market. Of course, leveraging the huge capital investment in the infrastructure can be an advantage for alternative fuels and not an impediment. This assumes that the alternative fuels do not require costly modifications to the infrastructure or cause unforeseen problems that interrupt the current supply of transportation fuels. For example, ethanol is more corrosive than gasoline and absorbs water, so gasoline and ethanol are not blended at the refinery but at distribution terminals downstream. This reduces the risk of corrosion in gasoline pipelines and maintains quality for the consumer. However, this approach increases cost of transport for ethanol to the terminal, since a separate distribution system, such as a dedicated pipeline or truck or barge transport, is required for the ethanol.

Ethanol has been successfully integrated into the transportation fuel distribution infrastructure because, as an oxygenated blendstock, it improves gasoline octane rating and reduces tailpipe emissions. Advanced biofuels are touted as drop-in hydrocarbon replacements; so, in principle, minimal infrastructure modifications should be required, assuming they meet all the existing standard transportation fuel specifications. Therefore, advanced biofuel integration in distribution infrastructure is expected, but integration with the petroleum refining industry or other industrial integration opportunities is still an open question.

The modeled capital cost for advanced biofuels processes poses a significant financial risk for developers who need to seek investments to build first-of-a-kind facilities. One strategy to keep capital investment to a minimum is to scale the process so that it is just large enough to make a profit. Larger capital cost reductions can be realized by co-locating biofuels facilities with exiting industrial processes to leverage existing capital investments and reduce operating costs by sharing utilities. Two critical aspects of co-location are matching the scales of the two processes and minimizing or eliminating the risk that the developing biofuels technology may adversely affect the already optimized process.

Direct biomass liquefaction technologies like fast pyrolysis, catalytic fast pyrolysis, hydrothermal liquefaction, and hydropyrolysis all produce a liquid hydrocarbon intermediate that one might naturally assume could be upgraded in a petroleum refinery.[6] In direct liquefaction, a significant capital saving can be realized by co-processing bio-crude in a petroleum refinery to avoid having to build a separate hydroprocessing unit. Hydrogen for upgrading could also be provided by the refinery, thus avoiding the additional expenditure for a steam reforming unit. In principle, the capital cost for direct liquefaction can be reduced by 30% if hydroprocessing is done in an existing petroleum refinery,[7] but other factors should also be considered. Most petroleum refineries are not close to biomass resources, so feedstock logistics must enter the calculation. If it is economically feasible to transport biomass to a biorefinery co-located with a petroleum refinery, there also needs to be enough space available to store biomass.

A common question asked of refiners in workshops convened to discuss integration is "How much oxygen can be in the bio-crude if it is to be co-processed in a refinery?" The short answer was invariably "none," but it is well known that bio-crude is not petroleum crude. The physical and chemical differences have been highlighted. Processes developed for crude oil require modifications to process conditions, and sometimes catalysts, to upgrade bio-derived liquids. In fact, some bio-oils and bio-crudes are not miscible with petroleum fractions, so co-processing is not practical.

Corrosion is another concern, because bio-crudes have low pH and may not be compatible with existing materials of construction. All else being equal, it is important to address the technical and economic risks associated with bio-crude co-processing in existing refineries. Modeled 2000 tonnes per day biorefineries are projected to produce approximately 3000 barrels of bio-crude per day; existing refineries process 100 times more, with the largest US refinery processing over 500,000. One co-located biorefinery would add only 1% or less of the total volumetric capacity in an existing refinery. The risk associated with adding even small volumes of bio-crude needs to be extremely low so as not to upset or interrupt normal refinery operations.

Another integration opportunity comes at the front end of the process with respect to the feedstock. Many existing industrial processes have residue streams that could be used. As discussed above, some of these residues are considered low quality, but matching the resource availability with the appropriate biorefinery scale may have economic benefits. The feedstock material and logistics costs would be greatly reduced compared to stand-alone biorefineries, and reduced equipment costs for feedstock preparation could be realized in favorable integration scenarios. One example is the co-location of a biorefinery with a pulp and paper mill. Hundreds of tons per day of softwood logs and chips are delivered to the mill, where cellulose is separated and turned into paper products. The bark from the logs can be converted into biofuels if the capital and operating costs of the co-located biorefinery are competitive with the current use for bark in a boiler for steam production. Hemicellulose can also be removed during the pulping process and converted to biofuel as long as this does not have a detrimental impact on pulp yield and quality.

Integration opportunities exist across the biofuels value chain from feedstock to upgrading to distribution. The challenge is to identify those opportunities that reduce the technical and economic risk to the existing infrastructure and improve the economic competitiveness of advanced biofuels. Win-win scenarios that positively affect both the existing industry and the developing technologies will obviously be the most successful.

Got Products?

One of the distinguishing chemical characteristics of biomass compared to fossil fuels is the high oxygen content. Most biomass conversion processes are optimized for removing the oxygen as water, CO, or CO_2. The technical targets for advanced technologies include efficient utilization of hydrogen with maximum carbon efficiency to produce hydrocarbon-rich intermediates and biofuels. However, biomass is composed of three structural biopolymers—cellulose, hemicellulose, and lignin—that contain significant oxygen functionality. Capitalizing on this functionality to recover high value products that are difficult to synthesize from other feedstocks, like petroleum, is an evolving strategy for improving the economic feasibility of biomass conversion while meeting the increasing demand for more sustainable and environmentally responsible consumer products. One example is lignocellulosic ethanol, where an oxygen atom from the carbohydrates is retained in the alcohol product during fermentation. Ethanol is blended with gasoline to enhance octane and reduce emissions. Other biochemical conversion processes are being developed and engineered with high selectivity to produce a variety of oxygenated bio-products, as discussed in Chapter 9. High volume biofuel blendstocks such as ethanol and other oxygenated fuel additives can be integrated seamlessly into the transportation fuel market without much risk. For high value commodity chemicals and building blocks, the markets must be able to absorb the additional product volume without saturating it, while providing enough room for market growth and expansion, so that the bio-products remain profitable. In general, high volume means low cost and high value means low volume. Chapter 11 provided a detailed discussion about opportunities for bio-based products, chemicals, and materials. A recent review highlights the opportunities for producing bio-based chemicals and polymers from lignocellulosic biomass.[8]

It's All About the Benjamins

The question is not whether we can make gasoline from wood chips. We can, by several methods: biomass gasification to produce syngas that can be converted to methanol that is dehydrated to produce dimethyl ether and then converted to gasoline range hydrocarbons (TIGAS process); biomass gasification to produce syngas that is converted to hydrocarbons via Fischer-Tropsch synthesis; biomass pyrolysis to produce bio-crude that can be hydroprocessed to gasoline and diesel range hydrocarbons; and lignocellulosic sugar fermentation to produce alcohols (ethanol and butanol) that can be dehydrated and converted to hydrocarbons via condensation and isomerization reactions. The real question is whether we can make gasoline from wood chips at a price that is competitive with current transportation fuels. Fuel cells are analogous to biofuels because it is technically feasible to produce electricity from fuel cells. In fact, alkaline fuel cells were used to power early Apollo space missions in the 1960s. Several types of fuel cells exist: alkaline, phosphoric acid, proton exchange membrane, direct methanol, molten carbonate, and solid oxide. But fuel cell technology cannot currently be used to produce electricity that is cost competitive with utility scale natural gas combined cycle power plants or renewable sources such as wind and solar.

Price competitiveness can be realized in several ways. First, continued research, development, and demonstration results in technical improvements that reduce the cost of biofuels production. Second, market conditions change to where the cost of conventional transportation fuels exceeds that of the state-of-technology advanced biofuels. Third, the environmental and societal benefits of biofuels outweigh the cost, so policies are enacted to subsidize biofuels. Fourth, higher value products such as chemicals can be co-produced. No one can accurately predict the future direction of the conventional transportation fuel market (that is the subject of another book), and policies and subsidies are strongly dependent on the politics of the present, which cannot be extrapolated accurately to the long term. Therefore, the most direct and permanent approach to reducing the cost of biofuels production is by driving innovation to achieve advanced technology.

But what is the real cost of advanced biofuels production? Although initial deployment of commercial scale plants has been achieved, operation has been suspended on a number of these due to unfavorable economics of operation. Technoeconomic analyses of empirically optimized n^{th} plant designs do exist in the scientific literature, and the rigor of these analyses has improved to where the uncertainty of the estimates has been greatly reduced. However, with the limited data on costs from commercial plants, uncertainty in capital and operating cost estimates is a financial risk that challenges investments in new technologies.

Most of the challenges discussed in this chapter—using lower cost feedstocks, integrating technologies, leveraging existing infrastructure, and co-production of fuels and chemicals—are directly or indirectly related to reducing the cost of biofuels. Technology advancements can lead to optimized processes with maximum yields to maximize revenue, but opportunities across the value chain can also improve the economics. This systems-level approach will be critical for developing creative and innovative solutions that propel the advanced biofuels industry forward.

The Heat Is On: Improving the Environment and Reining in Climate Change

The adverse environmental impacts of fossil fuel consumption for transportation have been well known since the early days of the automobile. The Clean Air Act of 1970 was enacted to reduce the environmental and health effects of particulate matter, ground level ozone, carbon monoxide, sulfur oxides, nitrogen oxides (NO_x), and lead. Significant strides have been made in reducing emissions in the transportation sector as cleaner burning fuels were developed and computer-controlled engines with optimized combustion led to higher fuel efficiency. The opportunities for co-optimizing fuels and engines in future was discussed in detail in Chapter 9.

The environmental and health effects associated with the criteria pollutants are short term in contrast to the impact of CO_2 emissions on global climate change. CO_2 is the main product of fuel combustion. Reducing CO_2 will require less fossil fuel use and greater utilization of renewable energy like wind and solar power and sustainable biofuels. Another option is to capture and sequester the carbon that is released from fossil fuel combustion. Carbon capture and storage technologies are being developed for stationary power sources, but capturing CO_2 in the transportation sector poses a much greater challenge. Electric vehicles offer zero-emission transportation if powerplant emissions are controlled. Replacing conventional fossil fuels with more sustainable, renewable fuels is an important option for reducing greenhouse gas emissions from transportation. Blending 10% ethanol in gasoline and 5–20% biodiesel in petroleum diesel directly displaces fossil fuels with biofuels. The Renewable Fuels Standard sets volume targets for fossil fuel displacement with up to 36 million gallons of first generation and advanced biofuels by 2020 to reduce greenhouse gas emissions by an estimated 100 million tons per year. The commercial aviation industry has ambitious goals to displace 50% of all aviation fuel with alternative jet fuels to stabilize fuel prices and reduce greenhouse gas emissions, as discussed in Chapter 10.

Reducing CO_2 emissions to limit the global temperature increase to <2 °C compared to pre-industrial levels is a daunting challenge that will require much more disruptive approaches. In fact, carbon negative solutions are needed to reverse the effects of global climate change by slowing and reducing the accumulation of CO_2 in the atmosphere. Chapter 13 discussed current options for sequestering CO_2 in long-life materials, but future technology is needed for utilizing CO_2 as a reagent for fuels and chemicals production. In the end, all options for reducing greenhouse gas emissions to mitigate global climate change will be needed to successfully transition to a carbon-neutral society.

Final Thoughts

Second-generation biofuels technologies have advanced to the point where commercial demonstrations have been deployed, and under the right economic conditions and policy environment these can continue. Advancements have improved process efficiencies and yields, but improvements need to be demonstrated at commercial scale, and capital costs need to be minimized to mitigate the financial risk to investors. The scale of these systems should be determined based on the resource availability for a given scenario so that cost and revenue are optimized in concert.

In developed countries, new biofuels technologies need to integrate with the existing transportation fuel infrastructure. However, in the developing world without that infrastructure, other options may be more appropriate. The economic paradigms that apply domestically may not be as restrictive in developing countries. This could open new opportunities for technology deployment to provide operational experience and identify future improvements.

Most of the easy options have been well considered. The main challenge is to figure out what <u>should</u> be done without settling for what <u>can</u> be done. The focus needs to be broader than just a new catalyst or incremental improvement to unit operations. Consideration must be given to the entire value chain to provide innovative technical solutions that are economically viable and environmentally sustainable, even for unconventional or niche scenarios. In addition, future biofuels policies should be mindful of protecting food security and land use for sustainability as advanced biofuels technologies mature. Pressure on biomass availability may arise as more biorefineries are built and compete for the same feedstock resources. Market dynamics will dictate future feedstock costs, but it will be critical to sustainably manage biomass supply for maximizing the greenhouse gas emissions reduction potential of developing technologies.

References

1. U.S. Energy Information Agency. Monthly Energy Review. July 2018, https://www.eia.gov/totalenergy/data/monthly/; 2018. Accessed 12 June 2019.
2. Hajjari M, Tabatabaei M, Aghbashlo M, Ghanavati H. A review on the prospects of sustainable biodiesel production: a global scenario with an emphasis on waste-oil biodiesel utilization. *Renew Sust Energ Rev*. 2017;72:445–464.
3. Schipper M, SBowsers R, Mayes F, et al. *Direct Federal Financial Interventions and Subsidies in Energy in Fiscal Year 2016*. April 2018.

4. Perlack RD, Eaton LM, Turhollow Jr AF, et al. *US billion-ton update: biomass supply for a bioenergy and bioproducts industry*. 2011.
5. Geissdoerfer M, Savaget P, Bocken NMP, Hultink EJ. The circular economy—a new sustainability paradigm? *J Clean Prod*. 2017;143:757–768.
6. Talmadge MS, Baldwin RM, Biddy MJ, et al. A perspective on oxygenated species in the refinery integration of pyrolysis oil. *Green Chem*. 2014;16(2):407–453.
7. Wright MM, Daugaard DE, Satrio JA, Brown RC. Techno-economic analysis of biomass fast pyrolysis to transportation fuels. *Fuel*. 2010;89:S2–S10.
8. Isikgor FH, Becer CR. Lignocellulosic biomass: a sustainable platform for the production of bio-based chemicals and polymers. *Polym Chem*. 2015;6(25):4497–4559.

Index

Note: Page numbers followed by *f* indicate figures, *t* indicate tables, and *b* indicate boxes.

C

T

U

V

W